Supported by the generosity of
Dan Ayling, Paul Bassett, Dawn Bazely,
April Chamberlain, Chris Fremantle, Claire Moran
and anonymous

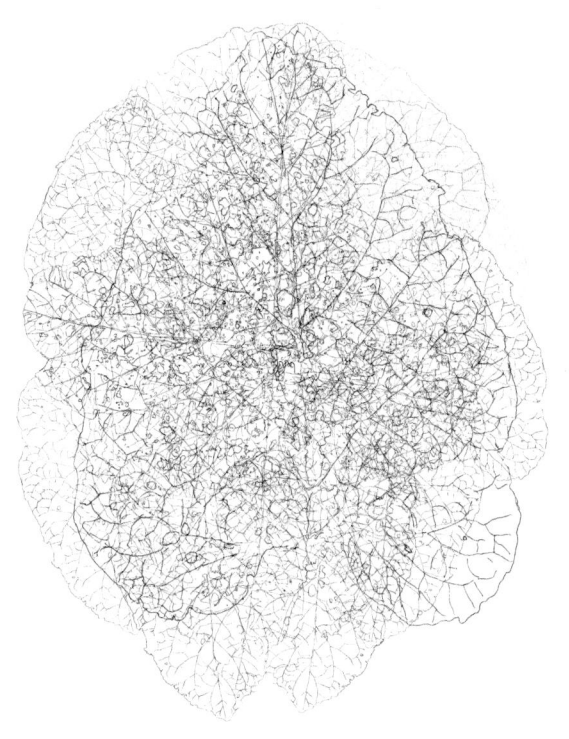

Brain Leaf

Nathalie Lavoie first drew in pencil on paper the venation of plant leaves in her environment, then digitized, multiplied, superimposed and organized these lines using digital processes to create new and complex structures. From this creative process, both systemic and intuitive, emerged a symmetrical and interconnected organization of brain structures. Moreover, it is these same mental structures which created these drawings. From this perspective of hybridization, humanity could learn to better understand the relationships that inhabit and connect us. Therefore, mind and nature would be more consciously linked in constant reciprocal action.

Nathalie Lavoie

becoming–Botanical
a post-modern liber herbalis

Edited by
Josh Armstrong & Alexandra Lakind

Published by   Objet-a Creative Studio, Glasgow, UK — 2019
                  SCIO registered in Scotland: SC048214
                  www.objeta.org

Editors   Josh Armstrong
                  Alexandra Lakind

Copy Editor   Gabrielle Kelenyi

Supported by   Royal Conservatoire of Scotland
                  University of Wisconsin-Madison
                  University of California, Santa Cruz

Printed   Mixam, United Kingdom, 2019

ISBN   978-1-9161283-0-9

Contents

EDITORS' NOTE

As the editors of this book, we are interested in exploring the reciprocal and responsive reaction to the act of *becoming-*Botanical. A *becoming* is always molecular and exists in the liminal proximities of bodies.[1] In physically consuming or absorbing another body, this botanical *becoming* exists within undeniably molecular micro-proximities. Through the act of ingesting, anointing or adorning oneself with, or growing up alongside a plant, we are partaking in an act of undoing ourselves and the plant—a breaking down of borders between the distinctive barriers of identity. The human withdraws as the botanical rises-up. These acts are not relegated to the anatomical or chemical but create rhizomes with language, history, perception, belief, and imagination. Thus, we are interested in exploring the complexities within current human|plant relationships, particularly the ways these relationships have developed within geopolitical, socio-economic post-modern times. This includes exploring the ways in which plants have been historically entangled in the terraforming practices of colonialism and capitalism as well as the deep histories of indigenous practices and knowledge systems (and how the former has been utilized against the latter).

In June 2018, we published a call-out for contributions from anyone: herbalists, scholars, researchers, historians, scientists, poets, naturalists, artists, activists. From nearly 100 submissions, we have distilled the entries to what is found in this publication. The process of editing and publishing *becoming-Botanical* has been one of discovering a global community of practitioners interested in working across ~~boarders~~ borders to intentionally cultivate a sustainable future. The entries in this book prompt us to ask: In these moments of *becoming-*Botanical, how are we—as humans—altered by our relationship with botanicals, and how are we altering the plants from which botanicals are extracted? What new knowledge, action, or experience can be fostered by shifting our perspective from the othering of nature to one of incorporation and companionship? How might these relationships foster ecologies of resistance and refuge and work towards envisioning radical and sustainable future relationships?

We chose to format this as a Book of Herbs (or Herbal): a published collection of entries describing the medicinal or cosmetic applications of plants and fungi along with a phyiscal description or illustration for identification. The existence of Herbals has been traced as far back as 2700 BCE.[2] While seen primarily as medicinal and scientific manuscripts, these ancient Herbals contain esoteric, mythical, and plant-based lore as well as information pertaining to the herb-gatherers or rhizomatists. An Herbal does not simply identify plants and their applications but provides insight into the complex relationships between humans and the various botanical ecosystems they foster and (re)produce.

The entries are ordered alphabetically by the plants' and fungi's binomial nomenclatures. Acknowledging the complexity of scientific colonial 'discovery' and cataloging of plants, we have asked each contributor to include a name for the plant or fungi in the language most relevant to the context of the entry. Likewise, authors have provided their own titles, affiliations, images, and citation formats and have written in their preferred dialect of English. Thus, you will see this book oscillate between various spellings and formats.

We are grateful to everyone who contributed and supported this project and honored to be in the company of so many interesting people and plants. While structured as an encyclopedia, we encourage you to approach this book as a rhizome, or complex intertextual root system: enter through any opening and move in any way you choose.

Josh Armstrong & Alexandra Lakind

F O R E W O R D :

Shifting Nature/Culture Relations in the Anthropocene

I grew up amongst the lively hustling and bustling of a small-scale avocado and tangelo grove in the hills of Valley Center, California. I spent my days following dogs as they roamed through layers of rotting mulch under the low canopies; I soaked in the sun with lizards laying on warm terracotta tiles; and I wandered from place to place as busily as the honey bees zooming from blossom to blossom. My grandfather planned and planted all the trees, and with them he planted his hopes and dreams for the future. The Grove was a refuge; a place to go when the surges and storms of life left you displaced, disheartened, or deserted. It was a place that gathered, a place that celebrated (weddings, birthdays, holidays), and a place that grew in dynamic ways.

What I remember most is the smell. The smell of ripe and rotting tangelos. Sunscreen. Manure. The stench of labor, life, and death. I watched as it became harder and harder for my family to sustain the grove: as water became increasingly scarce and more expensive, and large industrial farms took over the market. I watched as the Grove was sold and half the trees cut down. The trees that my grandfather planted, with gleaming pearls of pesticides that caused his cancer and death. Before we knew the detrimental effects of carcinogenic pesticides.

This story is personal. Intimate. Particular. Yet it is not unique. It's a story shared fully or partially by many small-scale farmers across the world. Representative of the bitter sweet

Poison Pearls

My father's death
a chemical cover up.

Poisons in the environment
spread on the fields,
hanging from trees
like the poisoned apple in
    Snow White.

Death disguised
in blemish free fruit,
bare hands spreading pearls
of cancer,
bleeding into the earth
sucked up into hungry mouths.

The sweet orange flesh of the
    tangelo
and the ripe green mash of an
    avocado
held more than sunshine.
It held sadness and sorrow,
pain from an I.V. line,
radiation that followed
memories of better days,
days of a trust that the rich
    brown earth
rubbed between our fingers
    brought life
and not death.

They knew, we didn't
we knew, they didn't
what they did tell us,
we ignored.

At what price do we lose?
Have we already lost the sweet
    orange flesh
of our future?

Do we still choose
the blemish free fruit,
even though we know?
Hindsight has no use
when we are blind.

The excuses build up like the
    toxins hidden,
a chemical cover up
my father's death.

Lauriel Adsit

necropolitics that initiated and sustain the Anthropocene—our new epoch of geologic time dominated by human impact. These post-modern Anthropogenic times are complex, contradictory, and multiple. This collection of essays, poems, and provocations is an exploration of the shifting relations—biological, social, technological—between nature and culture from an ontological, epistemological and ethical stance. They embody small acts of care and resistance towards addressing the urgent, multifarious, and world-historical moment we find ourselves in, centered on interlinked socio-political, economic, environmental, and spiritual crises. This collection follows a long tradition of nature/culture literature that records the complex relationships between humans and the various botanical ecologies they've coevolved with. Such records and representations help us navigate the ways these relationships have developed and changed. By imagining this book of herbs (or Herbal) as a diffraction grating, an apparatus that diffracts and disperses scholars and artists to explore a multiplicity of issues and areas raised by shifting nature/culture relations in the Anthropocene, it provides a beautiful kaleidoscopic view of the bitter sweet troubled times we find ourselves in. As you read this collection of essays and poetic provocations, I hope that you will let them ruminate until you can smell the sharp sweet scent of peppermint, taste the lavender honey pie, see the erotic dance of death and resurrection by the *Mimosa pudica*, feel the flowery seductions and sensations of blooming, and grieve the loss of our old growth tree elders. These stories are personal. Intimate. Particular. Yet they are stories that we all share. Partially connected and fully entangled. They must be told and held true. The Anthropocene functions as a post-truth phenomena. A capitalist cover up. Leaving us displaced, disheartened, and deserted without refuge. Whether we know it or not. This book gathers us through various provocations and creative ecologies of resistance and refuge, working towards envisioning radical and sustainable future relationships. Such relations require grappling with the inherited nature of our entangled pasts/presents/futures epistemologically, ontologically, and ethically so that the hopes and dreams we plan and plant have a chance to grow.

So, lather up the sunscreen. We have work to do.

Chessa Adsit-Morris
Assistant Director, Centre for Creative Ecology,
University of California, Santa Cruz

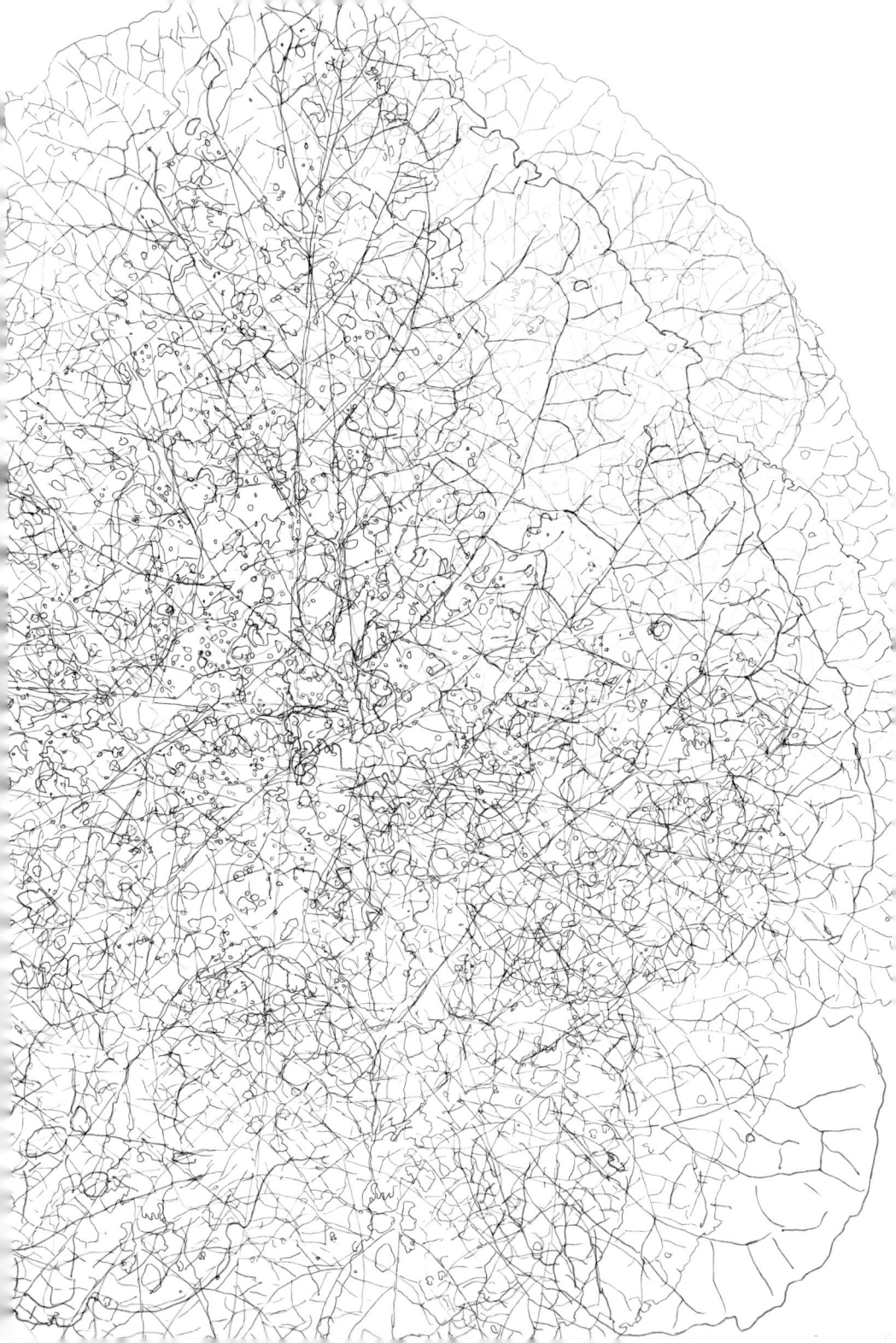

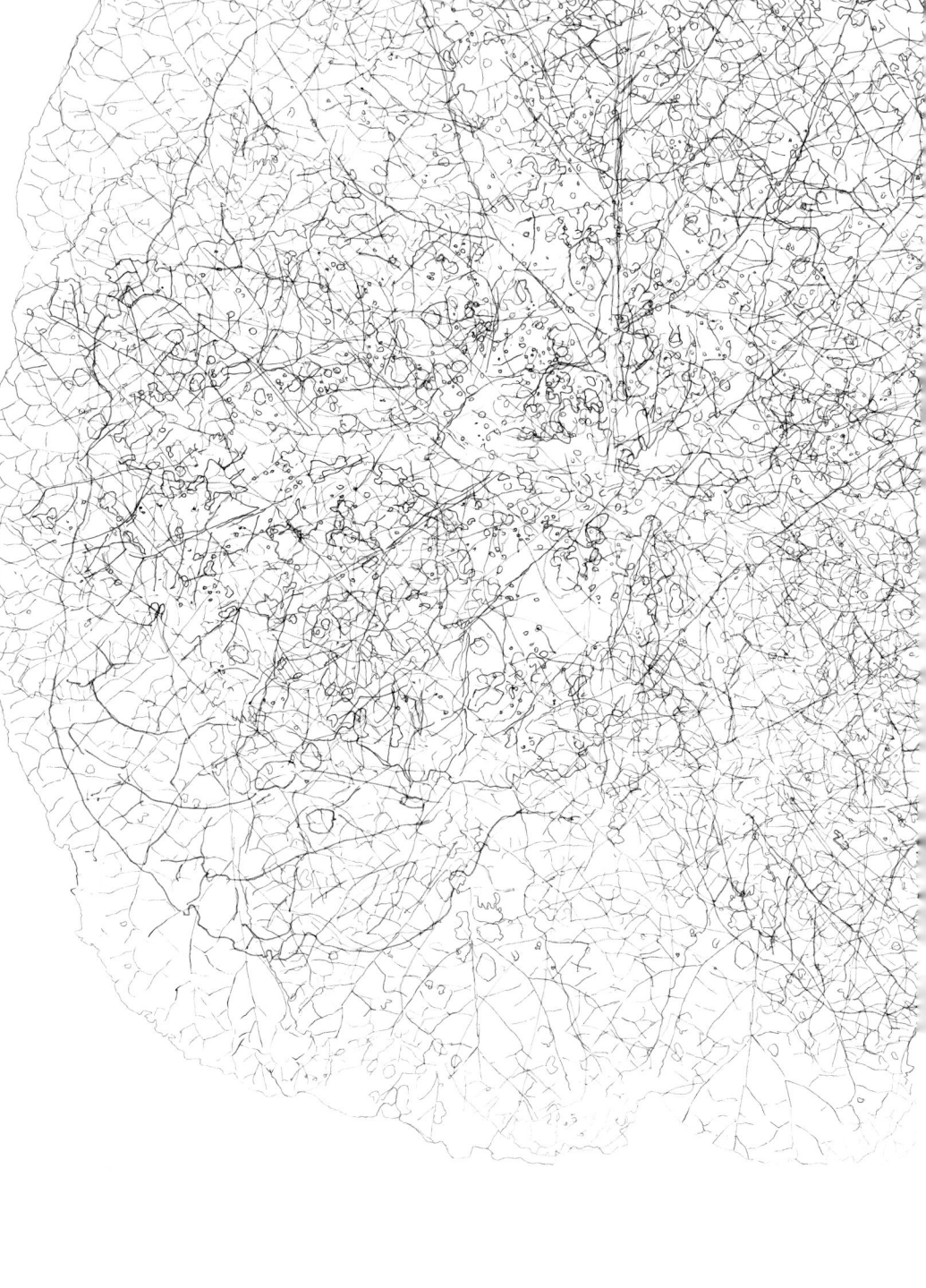

ALLIUM
CEPA

Onion

Waste Not, Want Not: Utilizing All of An Onion

It is deeply rooted in Lithuanian culture that something should be used in its entirety, whether manmade or natural, either by the original obtainer or passed along to kin. Possibly stemming from a combination of the harsh climate, short growing season, and terrible recent history of the brutal Russian and German occupations, completely utilizing an object was, until recently, a necessity. I hope this persists within the Lithuanian psyche during this rapidly changing, consumerist time. One such example of a natural substance used in its entirety is the onion, a crop that grows well in the cold climate and stays fresh through storage in the prevalent cool, dark cellars. Walking through Tymo Turgis—the weekly organic farmers' market in Vilnius—one will find farmers, with hands still muddied from that morning's work, delivering brown and red-skinned onions to shoppers. Though the potato is king in Lithuania, onions are ubiquitously used as a spice and flavour in an array of dishes—anything from pickles to roast chicken, borscht, and sorrel soup. But instead of discarding the brown onion skins after peeling, they are saved and utilized as a natural dye. Easter—an important holiday in Lithuania—is celebrated with family feasts and eggs dyed with natural colours and nature-inspired designs. To dye eggs, they are wrapped in discarded, dark coloured onion skins and boiled until hard to obtain a smooth, fawn brown stain. Aside from dying eggs, onion skins make an effective hair tint for those with blonde to light brown hair. To create a hair dye from onion skins, they must be boiled and the resulting liquid should be left overnight to further extract the pigments. After straining the onion skins out of the solution, the liquid can be applied to the hair, left to sit, and rinsed out to obtain a soft brown tint.

Elizabeth Georgian, Ph.D.
Science communication professional

ALOE   VERA

Aloe vera

Aloe, it is not from here. Nowhere I have lived is where it lives. *Vera*
    is a woman's name meaning true. Vera is dutiful, reserved,
    and has a sleek chiffon lap, low-heeled pumps, a soothing palm.
My skin is not the same skin as when I was a child, and yet it burns and
    burns: remember, it says, that night of summer vacation at age
    nine, lying in the little bedroom in Laguna Beach after a long day
    of salt sun salt sun salt sun, the air so hot it was white, everything
    seemed white in the froth at the shore, how it flamed, my body
    stabbing red and sleepless under the scratchy sheet next to the
    window, oleander breathing fruity silk in the dark outside. My
    whole self hurt, and it was a throbbing fever, an exquisite ache
    of pleasure at being scorched by life, to which I awoke in that
    moment ecstatic. The first time. I listened to the ocean rustling,
    surging at the crag, and felt such a comfort in the soreness of
    my skin which is everything, for without it we would just be open.
So that was the 1970s. Emphatically tan shoulders shredding
    to naïve pink. Now we know about Prevention First and
    Healthy Choices and *Aloe*. Now *Aloe vera* is everywhere
    because in the 21st century so many of the old ways are for sale.
Clean cool slime dripping from the green *Aloe*-shaped bottle,
    straight from the plastic leaf, slick as lube smearing where
    you want it wet, squish it in, rub the lectins around on you so
    they can recognize your dermal need, soft as a dog's tongue
    licking, crooning to the world's biggest organ, okay, it's okay.
If, on the other hand, you have an actual *Aloe*, you can slice a
    rubbery arm off and its wound will heal because *Aloe* is a

witch doctor—a genuine stranger, its odor musky as a field
animal or a bit of rot, velvety pointed and specially marked
with regular teeth, touch me touch me not like a sexy prickly teen.
I saw a film of a man using an *Aloe vera* leaf to masturbate, his cock
riding the jelly up the pale tube, and what was invisible were
the *Aloe*'s proteins seeking out the glycoproteins of the man's
cell membranes, to bind with them. What is and is not medicine?
Let me pour some sugar on you, says the potted *Aloe* to my tender
chest flecked with Wisconsin lake sand. Broken to pulp, gaping
and oozing its sticky insides like some saint. Let me pour it down.

Cynthia Belmont
Professor of English, Northland College

AMANITA BISPORIGERA

Destroying Angel

The alabaster white, smooth-capped sporocarps of this fungus are found in association with oaks and other hardwoods in forests. Aspects of its morphology and ecology assume particular resonance in the current era. Four such traits strike us: first, *Amanita bisporigera* produces two spores per basidium (its spore-producing structure), suggesting each individual mates with itself; second, the spores of a developing mushroom are revealed when the universal and partial veils break—the remnants of the universal veil form a cup around the base of the mature mushroom while the partial veil leaves a ring around the stem; third, if eaten, this stunning fungus is deadly poisonous; and fourth, *Amanita bisporigera* is symbiotic with trees—hyphae, the branching filaments that make up the mycelium, grow intertwined with tree roots and exchange nitrogen for carbon.

In the post-modern moment, as a luminous presence in our shrinking woodlands, the destroying angel may represent political and environmental dangers we confront and the challenges they pose for individual and societal transformation. In its reproductive autonomy, the destroying angel figures completeness, needing no other of its kind for species survival into the future. Bearing broken veils, it glows as a spectre of apocalypse, a word whose etymological roots point to uncovering, unveiling. Yet, in its threat of death and in the demand for attention made by its striking beauty, the destroying angel may function like a wrathful Buddhist deity whose fearsomeness renders him a force of awakening and transformation, who overcomes the karmic obstacles of anger and hatred. Perhaps most fundamentally, the destroying angel, in its subterranean networks, demonstrates the often invisible interdependencies on which the flourishing of planetary life depends.

Emily Arthur
Associate Professor, Vilas Associate Award Winner, Department of Art
Faculty Affiliate of the Nelson Institute's Center for Culture, History, and Environment
University of Wisconsin-Madison

Lynn Keller
Martha Meier Renk-Bascom Professor of Poetry, Department of English
Director of the Nelson Institute's Center for Culture, History, and Environment
University of Wisconsin-Madison

Anne Pringle
Letters & Science Mary Herman Rubinstein Professor, Department of Botany
Faculty Affiliate of the Nelson Institute's Center for Culture, History, and Environment
University of Wisconsin-Madison

Emily Arthur, *Destroying Angel*, 2018 →
Screen print with silver leaf on distressed paper coated with wax, 15 x 11 inches.
Unique Print.
These two specimens, *Amanita bisporigera* and *Russula brevipes*, were found near each other in a Michigan forest.

Zuri Camille de Souza, *Amaranthus*, 2017
included in Distant Nature, a political herbarium harvested and assembled in Palestine.

AMARANTHUS

श्रावणी माठ *Shravani maath*
राजगिरा *Rajgira*
Tambdi-bhaji

Leaves salted with manganese, magnesium, iron and phosphorus[1]; grains laden with sacred amino acids and precious antioxidants fed by a long, rose-tinged taproot that methodically draws nutrients out of arid, hot soil to embed it into leaf-tissues—here is immortality in all its forms, as ancient grain and persistent pigweed; tall, annual herb and pseudocereal. Scarlet-purple, deep maroon inflorescences; stalks of sand-like ivory-husked seeds stand tall, catching light as they slowly sway in a warm breeze.

1519: An Aztec[2] deity sculpted out of amaranth seeds and honey is broken into pieces, distributed as a festive offering; the grain is outlawed by Spanish colonizers and the Catholic church in an attempt to prohibit Pagan rituals and weaken indigenous communities. Fields are destroyed and cultivation of the crop is met with severe and violent punishment.[3]

2018: The Goan minister of agriculture campaigns for Amaranth—locally known as Tambdi-bhaji—to be named the state-vegetable, an ode to the many dinner and lunch tables that bear bowls and plates of Amaranth leaves cut and stir-fried with coconut and spices, an eternal companion to fish curry and rice. He promises support for those intending to cultivate abandoned fields. There is a unanimous decision made by the Goan government to restart iron-ore extraction the same week.[4] These fields are empty because the streams and rivulets that once refreshed them have been drained at their sources, their springs often located near the mines in fertile and biodiverse forest-zones amongst the ridge of the Western Ghats. A cosmopolitan genus—here is a plant that finds itself in a multitude of forms: as Amaranth, contemporary superfood sold in neatly-packaged and labelled boxes; as Tambdi-bhaji, grown in dry, cracked earth on the southwestern coast of India; and as Rajgira, glossy pearl seeds eaten during days of fasting and abstinence. Its small red leaves bound into bunches with twine, sprinkled with water, sold on a small square of blue canvas in the evening market. This is a plant that resists the toxic brutalization of glyphosate and excess fertilizer on its tissues. Whilst herbicides attempt to stifle the plant's growth, Amaranth contains pre-coded material within its genetic structure that allows it to adapt to the chemical constraints placed onto its growing medium.

Zuri Camille de Souza

ARGANIA SPINOSA; SIDEROXYLON SPINOSUM; ARGANIA SIDEROXYLON

Argan

Argania is a tree endemic to the calcareous desert and semi-desert Anti-Atlas and Souss valley of southwestern Morocco. It belongs to the order of Ebenales and of the Sapotaceae family. It is the only northern representative of this tropical family and was described as a botanic relic witnessing an ancient extension of tropical vegetation in southwestern Maghreb.

Argania is, for the Amazigh people of the Anti-Atlas, a mythical and sacred tree: the father of all and the symbol of their life, resistance, and way through time. The tree was there well before the first man and welcomed all who were born or came to take refuge in its forests. The mythical aspect of this tree can be related to the golden fruit of the wonderful garden of the Hesperides evoked in Greek mythology. This legend states that Heracles, looking for the golden fruit, reached the country of Atlas, present-day Maghreb. He beseeched Titan Atlas to relieve him by bearing in his place the celestial vale. In exchange Atlas asked his daughters, the Hesperides, for the "golden and bitter" fruits for Heracles. It was long believed that these golden fruits were oranges, but oranges originating in China were not introduced into North-African ecosystems until the end of the

10th century. Indeed, if we take into consideration that the legendary *golden fruit* makes immortal or plays a role against aging, it is quite surprising to note that modern research on argan oil has discovered its anti-aging power. There are many Amazigh beliefs about rejuvenation, and especially the Atlantides (Tachelhit Amazigh women of the Atlas Mountains) believed in a legend of the Hellenes (Greeks) regarding the golden fruits of Hesperis (Sunset/Maghreb). Furthermore, the *Argania* trunk is similar to dragon scales, and its thorny branches are a natural ban against tearing the golden fruit, except for the Hesperides or Atlantides (daughters of Hesperis and Atlas) as reported in the legend.

Argania is a multipurpose tree. Each part of the tree is usable, a source of income for the Amazigh people and of nourishment for their livestock. The Amazigh Jews prepared the *Aman-n-Dounit* brandy by fermenting and distilling the pulp which consists of many carbohydrates. They also wore pierced argan nuts on their necks to ward against the evil eye. Amazigh people use argan oil for its nutritional as well as cosmetic virtues. In the traditional pharmacopoeia, argan oil has been used for its multiple properties. It is recommended in the treatment of juvenile acne, eczema, and to relieve the cutaneous manifestations of measles and varicella. It is recommended for cleaning and disinfecting wounds. It is also recommended to relieve colic of small children. Amazigh women have used it to massage their faces after their Hammams (traditional bathing) for a satiny and soft skin and to reduce their wrinkles. Nowadays argan oil is used to empower rural women to become agents of their own development by involving them in cooperatives to improve their financial and social positions: the mythical golden fruit makes the modern Atlantides/Pleiades shine again!

Houssine Soussi
University of Ibn Zohr

Crushing argan seeds for producing oil, a typically feminine job in Morocco

A R T E M I S I A
VULGARIS

Mugwort

You use the asphalt crack to slink the way through.
You expand and follow along the wall to find sources for
    existence.
Your presence makes me feel exposed.
You demand I only focus on you.
I listen to you and walk towards you.
You invite me to drop my guard.
I wonder how close we want to become.
You offer your young frond to me.
I take your gesture and sink down to the asphalt.
I read your memory. We touch.

You are remembered as the witch, the healer, the magician, the
    dreamer, the protector in ancient time,
and as the power, the strength, the freedom, the controller for
    women's reproductive rights,
and as the intruder, the citizen, the invader of the United States.
and now you are the official fence line plant, the official
    building collapse plant, the official superfund site plant,
and yes, I remember when I was falling in love with you I did
    not call you *Artemisia vulgaris* or Mugwort,
for me, my love, you are the community organizer, the rhizome
    biology scientist, the lover of human disturbance, the
    mirror of our toxic existence.

Your cooling scent sprawls over my head.
I see a layered abstract green shape.
You blow on me and are sculpting into a new position.
I back off and my bones are pressing into the asphalt.
You want my attention. Your force asks me to become a corpse
      flower that smells like death.
I sense your heavy weight through my nostril.
Your dance for me is a move with the wind.
I sneeze and decompose foreign particles into the air.
You persist to be part of it.
I sharpen my eyes and notice our intimate closeness.
You invite me for a micro-drama love affair.
I decide to abandon you. I blur my eyes.
You don't give up. Your vibrations are present in darkness.
We both move on.
I hunt for a new gesture.
You hunt for the sun.

My interaction with Mugwort took place at 1067 Pacific Street, an urban lot in Crown Heights, Brooklyn, NY.

Andrea Haenggi

Andrea Haenggi, *An interaction with Mugwort*

Mucgwyrt

The Making of an Incense Wyrd - Wyrt Blaed

A loving alchemy of aromatic weaving, plant breath.
Enter the woods of Wyrd, soft, yielding, inviting, a deep
    massage of the green calls in the wild.
The cauldron of possibility swirls, murmurs, takes form, enters us.

Lovely moves, pressures, rhythms, strokes: the spirits come in
    fast, almost instantly.
Vivid, visible as if eased out of the earth and on to centre stage.
We want to follow.
Breathing it in takes us there, to a wonderful liminal place.
Suddenly sucked into the plant yet still sitting in the studio,
    sucked through its branches
Nestled into an organic bed of magic.
A world beyond and somehow behind, inside.

Musky, deep, dangerous, thrilling, the promise of something
    deeper
Wants to give pleasure, wants to taste good, wants to make us
    happy, like a fabulous lover.

A guardian with sword and shield, a rough English sort of
    warrior,
This plant magic stirs that feeling in swirling, smoking
    rhythms.

"Lend us your voice of power."

Breathing the English landscape, Sussex under a full moon.
Wyrd, a thousand and a half years ago.
Smoke undulates, thick and moist, dancing before our faces
    like a Wyrm.
Writhing, dancing, snaking, teasing plumes engulf us.
Hot, green, wild.

A tantalising tapestry of ancient memory.
Plumes of recognition of our distant past, plunged back in to
    Anglo Saxon times.
The aroma of a great feast.
A dream being narrated in a voice like a drum beat, strange,
    compelling, unexpected.

How to be close to tribe, tribal rhythms so powerful they are
    frightening.
A celebration of intimacy of spirit. The world needs this.
The few, the many, we haven't understood how to deal with it.

Heat and smoke and crackle of fires, the love, the comfort.
The feeling of security.
Somehow its damp, smoky pleasurable green presence makes it
    hard to leave.
We want it for a companion in life, dancing, loving, touching,
    sharing.

The spirits of this plant come closer.
Very close. We sense the closeness in the way we do when in
    love,
Like transcending journeys of the soul, glorious waves of
    intimacy and emotion.

They enter our dreams, our bodies, our vision.
Our senses flooded, transported.
Gathered up to the heart of our tribal ancestors.
Gathering up for us the things we have done and tried to do.
Into the slow deep pulse of life, a way of being.

Riding tribal closeness with the protection of this fabulous
    plant nature magic.
But there is another dimension.
For it can ride us if we let it.
There lies change, hope, a way of being.

It can be touched, called upon, there is indeed another
    dimension here.
Otherworld threads dabbed on to the face of green.

Sweetness.

Inhaled, drunk, burnished into the body,
burned into writhing wyrm form, there is palpable sweetness.
The sweetness of secrets shared, clues hinted at, messages
    inhaled.
Healing, sacrament, celebration.
The scent of wine, meat, tribe, magic. Wyrd.

Enabler, partner, lover.
Watching the smoke form, we understood.
Experiencing Wyrt Blaed slowly, slowly we are changed,
    empowered, greened.

Lesley-Caron Veater[1]
Transpersonal Therapist
WindhorseDreamwork.com

Professor Brian Bates[2]
redskap@btinternet.com

Basidiomycota biocompositum

Mushroom® Materials by Ecovative Design

BASIDIOMYCOTA BIOCOMPOSITUM

Mushroom® Materials

Your legacy species erased through trade secret,
Your mycelium blends enjoy patent protection.
Ecovative innovated upon the world's oldest kinds,
Dank fungus,
Lowly mushroom.
Hybridized now with compatible botanicals,
You are pressed into service for higher purposes—
Homo sapiens' creative aspirations.

Mushroom® Materials, your proprietary name
Boasts solid-state substance on a plant-based substrate.
We can see
What you can be made into
On the company website's "Shop" page:
MycoFoam™ packaging for products,
Eco-friendly planters for plants,
mCore™ blocks and beams for building things,
Gummi bear statuettes for gardens, even—
Anything at all, really,
Which, when broken into morsels and
Left to elemental forces,
"will return to the earth
in about a month"
(your registered trademark notwithstanding).

The ultimate proof of concept,
You are feasible and then some,
Your durable disposability the very
Yin and yang of our yearnings,
We who aspire to make the world and
Hide the evidence of our making,
We of the curious dual drive to leave footprints,
Though they must not be too deep or too dark.

Your material properties,
Oh, mightier mushroom,
Tower over those of your
More pedestrian brethren,
Mere crop waste and yeast. You
Outcompete Polystyrene:
Denser, stronger, more resistant to heat.[1]
Though less likely to last, you
Present possibility.

Oh, mysterious not-quite-mushroom
Muse of mine,
Future of plastic,
One-day, too, maybe, eater of same,
Are you still a bit
Plant?
Have you taken a turn for the
Animal?
Or have you at last,
Through a mix of human and hyphae ingenuity,
Surpassed even your own exclusive kingdom?

Stephanie S. Turner
Professor of Rhetorics of Science, Technology, and Culture
Department of English, University of Wisconsin-Eau Claire

CARPOBROTUS GLAUCESCENS

Pigface (English)
Bubbracowie (Jandai,[1] Quandamooka Country,
Stradbroke and Moreton Islands)

Carpobrotus glaucescens is a medicinal and edible Australian plant that grows along the Southeast coast's sand dunes. With anaesthetic, antibacterial and antiseptic properties, it can be used for marine animal and insect stings, burns, and headaches, and its salty-sweet fruit can be eaten. The artwork* is from a body of work created for a visual art PhD on medicinal plants from Minjerribah/North Stradbroke Island, informed by consultations with Quandamooka community members.[2] By fusing plant and photographic materials and subjecting them to natural decomposition, the work encourages natural processes of decay and renewal, reveals the beauty in decomposition, and raises notions of its transformative cycles.

Renata Buziak, Ph.D.
Photo-media artist
Queensland College of Art, Griffith University

Artwork on p. 24-26

DALDINIA CONCENTRICA

Cramp Ball
King Alfred's Cake
Coal Fungus

The stooped ash was dying: fragile and brittle and slumped on the strong shoulders of the young in this ancient forest. The fallen leaves of last-year's fashion already veiled those fallen prey to time as the withering ash struggled on. Curious black pustules disfigured the aged tree, like the burnt remains of King Alfred's neglectful baking. Smooth and tough, these obsidian marbles clung tightly to the bark, devouring the remains of the once-proud veteran. It was a fungus I knew well, a saprotroph of hardwoods, the cramp ball.

Night was drawing in and the white flakes of early winter had begun to descend, drifting lazily down. Like a damp cloth tossed over my face, the chill was already settling in. I would need a fire tonight, I thought, fortunate that Mother Nature provided. A sharp tap, tap told me which of the fungi was too young, which was rotten, and which would greedily take a spark. All I would need to get an ember going. Next to come was that moment of prehistoric wonder, as light was to be brought forth from darkness.

Daldinia concentrica, named so for the silver and black concentric rings hidden within—it is this coal-like interior that made the cramp ball a tool in times long past: the fire-maker. Clink, clink, clink went the steel on the flint, throwing a meagre showering of white-hot sparks bouncing over the rings, sending them flashing and dancing like lightning against the storm clouds. The ember took hold, smouldering inconspicuously until I shepherded the heat with my breath, and smoke billowed as the ember erupted amongst my tinder bundle to start my simple camp fire. Heat and light from death and stone. Perhaps fire was once regarded as wizardry, or spirits called forth by a shaman. Perhaps it was a gift from the gods: light-bringer, murderer of shadows, and sacred ward against our own conjured nightmares. From knowledge passed down by word of mouth, our ancestors used this fungus for thousands of years—taking the gem of living-coal with them wherever they went. Perhaps it was a means to preserve the spirits of the dead, forever linked to the memory of the lost. Perhaps the black ball was a symbol of life and death: from the demise of the living the fungus feeds, recycling the energy and returning it to the world. Or perhaps a representation of mankind itself, of complexity beneath a simple surface, denoted by the web of hyphae burrowing deep.

> Today, inconsequential. Forgotten. Just a strange
> black blob on an unnamed tree. Has another of
> humanity's ancient secrets been lost?

Perhaps I will keep this cramp ball with me, I pondered whilst staring into my crackling fire, listening as the wood popped and hissed. Perhaps this could be my token of the mysterious ways of the wild, of how death brings life. Perhaps this simple fungus was my beacon of light in the darkness.

With thanks to Kate and Rachael

Bradley Fairclough
Ecologist

DELONIX REGIA

Flame Tree

Every other morning, I walk in Abuja's Millennium Park, a large green
space in the city centre where residents spend their leisure
time. I started walking because I wanted to reconnect to my
city, a place my family had lived for several decades. Though
I frequently oscillated away to other cities and continents for
work or school, Abuja was the place I called home. By walking
slowly, meditatively, and breathing in copious amounts of
fresh, fragrant air, my intent was to sprout roots, to ground,
reconnect, and earth myself to an environment. I felt somehow
orphaned from long stints away. I wanted my body to graft to the land.

Over time, as I walked off the main pathways towards wilder areas
on the parks edges, I began to perceive the park as a sort of
conscious benevolent entity. It was aware of my presence, as I
was aware of its aliveness. We would have silent conversations
that had less to do with words and more to do with mood and
resonance. I responded to its promptings by picking up the
organic and inorganic material it would present to me: a fiery
red petal would draw my eye, I would pick up a shiny black
seed just to feel its texture. I knew nothing of botany. I barely
knew the names of the flowers, seeds, and trees around me.
As the seasons changed from dry to wet, my senses were
engaged in the shifting, and we went through processes together.

Soon enough, torn up and barely visible photos of visitors left in the grass
among branches and fallen leaves began to reveal themselves.
The glossy paper provided a bit of protection from the rain, but
after a night left to the elements, the images were crumpled,

soiled, and stained. Slowly, I started to collect these fragmented, beautifully damaged and mysterious images gifted to me by the park. And so, we began to cross pollinate, to co-create hybrid children, photo-collage illustrations that refer to botanical drawings describing the nature, process, and descriptions of plant life. Who were the people in the images, and why would they tear up their images so consistently? Some photos were obviously bad prints, discarded when inks would fail to saturate the photographic paper accurately. Just like looking over the surface of a plant, it was impossible to know the impulses and inner workings of the people in the pictures. And maybe it didn't matter so much as they began new lives in the grass. I wanted to look through the surface, to peer inside and see the connections and relationships between the fragmented pieces and the nature around it. I also wondered what happened to the photographs left in the soil, the ones I failed to recover. Did the other plants recognise them as the same? Or did the plants sense the strangeness of the photographs as they refused to obey the laws of nature to sprout and grow?

Rahima Gambo
Visual artist and documentary photographer

Rahima Gambo, *A Walk* — Abuja, 2018 →
Organic & inorganic photo-collage-illustrations
(same on p. 32-33)

Delonix regia

Stem

Epidermis

Pith

D I F F E N A F I A

(endosymbiont, transgenic organism, hybrid)

DiffenAfia may look like a regular, common plant, but the truth is not that simple. *DiffenAfia* is a hybrid.

How it happened:
The idea was that if you can create a human-animal hybrid, you could also, in theory, make a human-plant hybrid.

I decided to add my DNA to a common house plant—*Dieffenbachia*. The condensed DNA collected previously in the WIT laboratory by Dr. John Phelan was dissolved in water. The solution was then introduced to the plant bit by bit, day by day with a drop bottle.

Is it new specie?
"She" did not behave as a regular plant.

Warning:
Direct contact may cause skin irritation and an irresistible desire to escape, to hide from the world...
This condition does not generally persist long. After the disappearance of the visual manifestations everything can return to normal.

This project posed the question, 'Is this still just a normal decorative plant or was its status changed?' It does include fragments of my (human) DNA, and it is also a living organism. This way, I am questioning the legal status of human-animal embryos as well as other mutated organisms stored in laboratories or 'semi-living' fragments isolated from their existence.

Agnieszka Krym

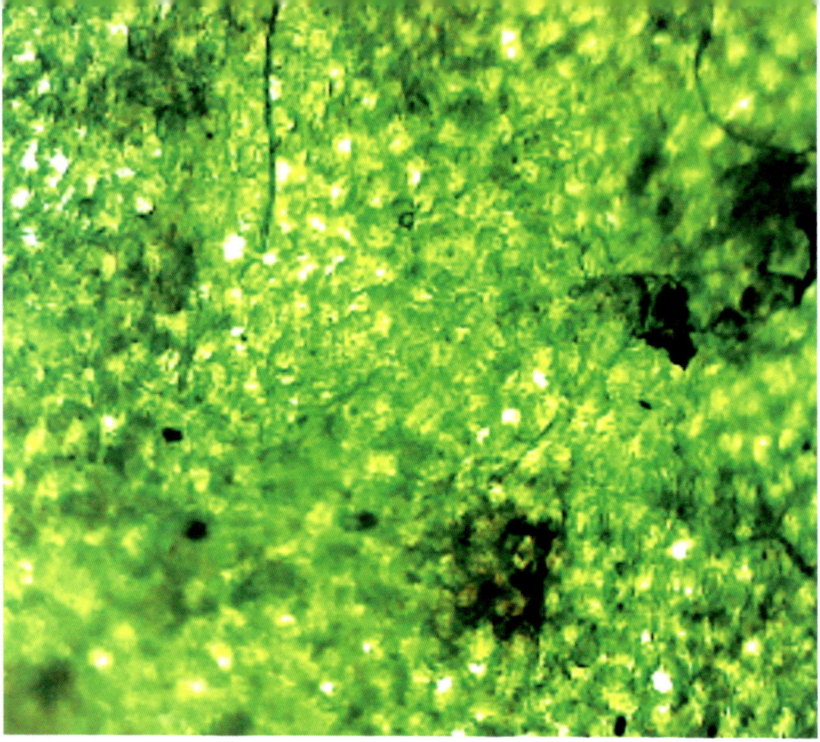

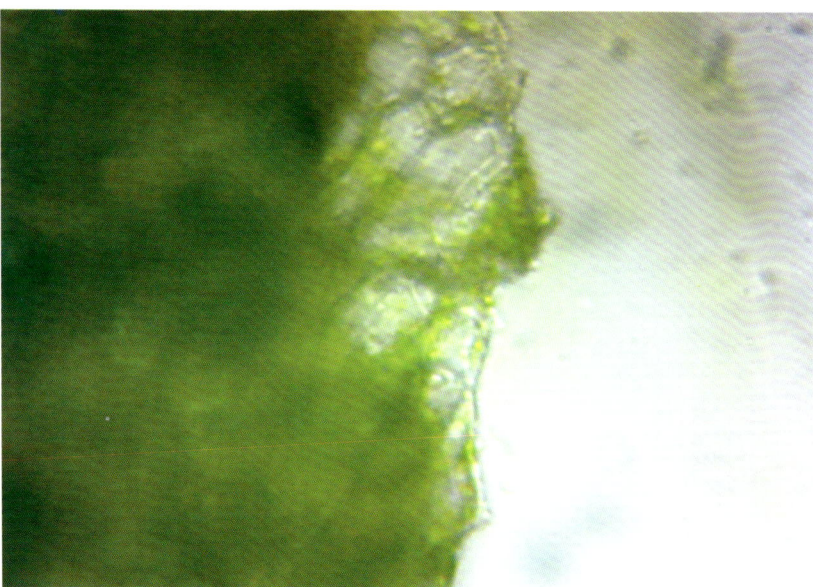

Agnieszka Krym, *DiffenAfia*, 2016
↑ plant section under microscope 2 years after introduction of human DNA.
↓ leaf section under microscope 2 years after introduction of human DNA.

EQUISETUM

Horesetail

It grew at the bottom of the valley, in the margins between river and
earth. Feathery, vascular, the horsetail, collected under the light
of the full moon, was a channel through which to harness the
energy of nature. I was a pagan. I was 16, and I lived in a tree
house at the Stanworth Valley anti-road protest near Preston
in the north of England. We collected the horsetail and ground
it into paste to put on our hair to make it strong. The ritual of
the horsetail shampoo represented both a rejection of the
capitalist culture I abhorred and a practice through which to
connect my body to the processes of Earth. The energy of the
horsetail, I thought, gave me power and agency when my neck
was d-locked to diggers in my impassioned attempt to stop the road.
Horsetail, described as a "living fossil," has not evolved since the days
that it carpeted the forested Earth in the Palaeozoic era. It is the
only surviving *Equisetum*: a plant that reproduces through the
distribution of spores. Horsetail thrives in the borders between
wet and dry land and has been used by humans throughout history
to heal wounds, to treat fluid retention, or, due to its high silica
content, as a useful aid for the maintenance of skin, hair, and nails.
The road was built. Yet, on a trip back to the ravaged Stanworth in
2010, I was glad to see life continuing beneath the motorway
extension. The horsetail was there at the bottom of the valley.
It stood out to me as an agential connection to the prehistoric
woodlands that once housed the creatures which would
become the fossilised remains that constitute oil. Somehow
the horsetail, the histories of carbon-fuelled capitalism

stretching back across the vastness of time, and the now ever-present thundering of overhead traffic seem inextricably entwined. I no longer throw myself under diggers in order to change the unstoppable juggernaut of capitalism. However, in this age, when the divide between nature and culture has fully broken down, I argue that small acts of care and resistance are imperative should we chose to be awake to the entanglement of which we are part. One such act can be found in the ritual of gathering horsetail on a moonlit night.

Freya Zinovieff
Simon Fraser University

G Y M N O C L A D U S DIOICUS

The Kentucky Coffeetree

In late May, along the streets of Chicago, you can find thousands of thick, brown, fleshy pods littering the sidewalks and oozing green goo. They are the fruit of the Kentucky Coffeetree, *Gymnocladus dioicus*. Tires of passing cars are the only thing that can break these sturdy fruits open, scattering their even more impenetrable seeds among cigarette butts and soda cans. For whom are they? These fruits seem to defy the standard botanical conceit of tempting animals to eat them, for not only are the seeds as hard as steel, but for most animals, ingesting the pods can lead to vomiting, muscle paralysis, convulsions, or even death.[1] Some biologists think the Coffeetree may be an "ecological anachronism" in which the natural animal partners that ate its fruit and dispersed its seeds are now extinct.[2,3] The characteristics that encourage specific animals to consume a plant is called its syndrome—seed number, fruit flavor, color, size, etc. The absurdly tough seeds of the Coffeetree—strong enough to withstand most crushing bites and toxic enough to deter all but the most massive of metabolisms from eating it— seem suited to mastodons, giant sloths, and other gargantuan mammals that roamed across North America until about 10,000 years ago. But a relative newcomer to the continent, a mammal migrant known as *Homo sapiens*, are thought to have hunted mastodons on a grand scale, perhaps to oblivion; the Coffeetree became a flora bereft of its mutualistic fauna. Or did it? After all, the newly arrived hominid mammals found novel uses for the potent plant. Native Americans supposedly threw its pods into ponds to stupefy and harvest fish. A wide range of uses were found across tribes: the Pawnee pulverized the Coffeetree's root bark into snuff to revive the comatose or alternatively to cause sneezing for headache relief; the Dakota used it with other ingredients to make black dye; the Omaha found that an infusion of the root (as an enema) was an "infallible remedy for constipation."[4] European colonists supposedly roasted the pods, neutralizing their poisonous alkaloids, and ground the seeds to make a coffee substitute. Even the government has a point of view: the United States Department of Agriculture rates the tree's wood as "good" for coals, "medium" in its ease of splitting, "good" in fragrance, "low" in smoke, and "few" in sparks. There are also the aesthetic properties of the tree: sight and sound. Some admire its spring leaves of "a striking pink-bronze color, turning to a dark bluish green above in summer" with golden yellow leaves in the fall; in the winter, "decorative clusters of the large pods rattling in the wind make for an exceptional winter ornamental."[5] The Missouri Botanical Garden holds a slightly different view about the tree's autumn yellow, calling it "undistinguished," although they agree that "mature female trees with hanging seedpods can be very attractive in outline against a winter sky."[6] Drought-resistant, tolerant to pollution, and happy in clay and poor soils, it is a natural choice for urban parks and has been used as a bio-remediator of mining waste lands. At once an ecological orphan and adoptee by way of human hands, the Coffeetree may not be so much an anachronism as an epitome—a quintessential plant of the Anthropocene.

Andrew S. Yang
School of the Art Institute of Chicago

Andrew S. Yang, *Novel Modes of Dispersal: A modest proposal for a new econo-ecology*, 2015
Kentucky Coffeetree Seeds, Gumball machine

IRRUPTION #1

Stop and Smell the Roses: a flower's guide to pleasure

Is it spring?
Find a group of people. The less you know about each other and the closer you live to one another the better.

Begin walking. Take the most familiar path.

Stop.

Confederate Jasmine—the smell of melancholy longing and midnight bike rides in waxy spirals and vortexes

Each time you see a flower.

Contemporary Rose—the smell of first kisses at pioneer camp and stained glass petals stuck in tea cups

Listen.

What do you hear?

Pineapple Sage
high pitched vibration
giggles
shrieking fairy-tale cartoon monsters

Come closer.
Toward an intimate proximity.

Describe what it looks like.

Gladiolus
a flamenco dancer
a monk's robe
contrasting colors
sharp and loose.
"hey sexy bees put your sexy parts in me"

Let go of the need to explain the flower through structure
and the Western lens of why and how it works.

Lantana—like confetti on a cupcake that you know is going to hurt
you when you eat it

Stay curious.

Magnolia Tree—Did you know if a tree flowers before it makes leaves
it is called a precocious tree?
Who came up with this?
Probably a man...

Come closer.
Make gentle contact.

Indian Arrow Root—the purple-red inside a flame of dragon's breath
with seeds like black beads on a necklace that will burn your tongue

Ask the flower for permission before you hold it in your hand.
What does it feel like?

Petunias
the cheek of an old withered woman

Naked Ladies (Amaryllis)
the cone is hypnotic
I want to stick my tongue in there

Come closer.
You are in make-out proximity now.

Heirloom Rose Bush—when Shakespeare wrote about the rose
he meant roses that looked like this: wild
a violent splash of color like an orgasm

 ...a male *orgasm*

What does the smell make *you* feel?

shock
a deep feeling that is familiar
a quivering in my groin
sticky sweat in the summer
a vibration in my third-eye crown chakra

If it takes two hours to walk two blocks then you are doing it right.

Slow down.
Ask permission.
Take pleasure in the process
and place no expectations on the outcome.

Meryl, Greg and Katya
from becoming-*Botanical Somatic Movement Workshops*
led by Meryl Murman

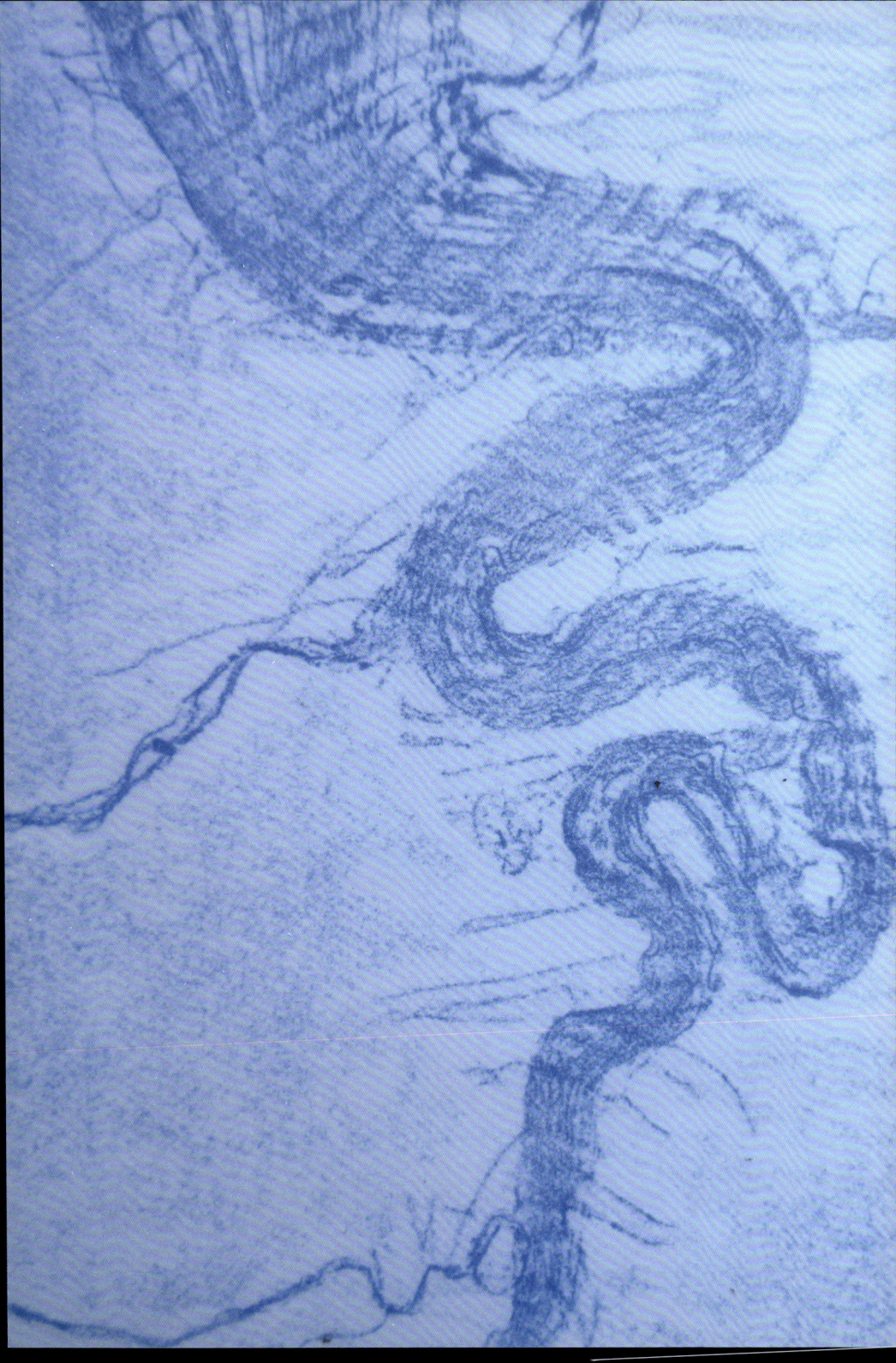

HELIANTHUS TUBEROSUS

Sunroot
Jerusalem artichoke

Champlain encountered Jerusalem artichoke (*Helianthus tuberosus*), or sunroot, while exploring Cape Cod in the early 17th Century and sent this artichoke-flavoured tuber back to France. From there the tubers dispersed throughout the world to be lauded, appreciated, maligned, and forgotten. Of all the names commonly encountered—such as Jerusalem artichoke, sunchoke, girasole, and topinambour—no Turtle Island Indigenous names, and few autochthonous entanglements, appear beyond oblique mention or passing reference.

Sunroot is both gourmet and privation food, alcoholic beverage and alternative fuel feedstock, invasive weed and saviour of the family farm, and an "underutilized resource." Treated as terra nullius, or "empty land," sunroot is simultaneously colonized and uncolonizable.

The sunroot, despite her myriad uses, remains peripheral. Her persistence at the margins of agricultural awareness, at field-edges, riverbanks, and at the limits of living memory is rooted in her resilience, wild fecundity, and canny partnerships with watercourses and other living things. A century ago, sunroot was recorded at the edges of cornfields and on the banks of rivers providing sustenance to Haudenosaunee husking parties. Some women who developed a taste for the tubers

← Basma Kavanagh, *Map Detail*, 2013
In its Own Tongue – Tools for Reading the River, Rabbit Square Books

were called artichoke eaters.[1] Onöndowa'ga:' chief Corbett Sundown recounted the origins of the Three Sisters: after Sky Woman's daughter died from giving birth to twins, corn grew from her breasts, squash from her navel, tobacco from her head, beans from her fingers, and sunroots from her feet.[2] Indigenous botanical knowledge is not organized in a book-form herbal. Instead, this wellness code is woven into the dynamic relationships with waterways and landscapes, seasonal cycles, traditional songs and stories, our flesh and the flesh of the more-than-human with which we exist. Reflecting on our sunroot entanglements compels me to ask: do old women, young women, girls still have the appetite to become 'artichoke-eaters'? Can men or two-spirit people become 'artichoke-eaters'? What appetites—or future horizons of being—could be engendered by renewing our relationships with sunroots? Neither here-nor-there (but both) and of the frontera...the borderland. Could sunroot reciprocity be a root/route or passage, "a hole out of the old boundaries of the self?"[3] My meditations on belonging to this continent return to the language of this land. Only by re/learning this language, embodied in the sunroot-mind—or mycelial-mind, cricket-mind, whale pod-mind—can we hope to revive the vital relationality necessary for the deep continental citizenship required to belong.

Keith Williams
Special Projects Advisor, First Nations Technical Institute
Doctoral Student, St. Francis Xavier University

LAURUS NOBILIS

Bay Laurel

A single bay leaf on the tongue
is enough to infuse
the pulp of your mouth
with flavour.

A single bay seed on the tongue
dissolves like a pill,
takes root
in the guts;
rhizoid hairs embed
in the wet flesh
of the lower intestine.

Epithelium tissue cracks
into fissures of bark;
dry leaves blister and bud,
erupting from the scalp, the tips
of the fingers,
the hard fork
of the pubic junction.

If the body is found to exist,
it will be stripped,
splintered,
tapped.
The skin will be burnt,
the flesh shaved
into flat shapes.

The hair and the nails
will be added to tomatoes,
stock,
fresh egg pasta.

Hannah Cooper-Smithson

LAVANDULA X INTERMEDIA

Lavandin

Next time you ride through an airport packed with endless race horses from the cosmetics industry, don't forget about me, the sturdy mule from the Lavender family. Don't forget how humble and calming my effect, put five drops of lavender oil on tissue, combine with a neck massage, and imagine flying through blooming purple flower fields to cope better with travel and stress.

Some elders of my family already helped Egyptian friends to balm their mummies, and Greek physician Dioscorides advised to drink me as tea infusion against chest complaints.

My family name sounds very much like the Latin word "lavare," which means to wash, and I was also used in ritual baths and sprinkled onto cleaned linen. My delicate scent banishes some insects, like moth and flea. Bees instead love to suckle the very sweet nectar of my flower spikes. Lavender honey is used in Lavender Honey Pie and this is like no other pie!

In between the over forty subspecies of my family I have a very special sister called *Lavandula angustifolia*, or true essential Lavender, the one with the narrow leaves that bear an even more delicate scent than mine. This special sister has higher levels of Lynalyl, and Linolool acetate. Lesser levels of these specific compounds are found in my own hybrid form because I am a natural crossing between *Lavandula angustifolia* and *Lavandula latifolia* becoming *Lavandula x intermedia*, or commonly Lavandin. I am more patient, hardy, and long-lived than my sisters and brothers. My higher levels of sharp and intensively smelling camphor compound and my faster growth make me currently the most cultivated and sold. Me, the sturdy soothing mule within the Lavender family.

Renowned French physician and the author of *Flora Gallica* Jean-Louis-Auguste Loiseleur-Deslongchamps was the first to mention my name. I think it's a very nice name for a mule. Only recently has science returned to research the medical properties of my anxiolytic calming effects and other benefits in treating dermal diseases, like fungi or infections. But I am very jealous, because my special sister, the one with narrow leaves, has an even higher medicinal potential, but maybe because she is so picky about soil and surroundings.

The European perfume industry is traditionally linked to the essential oil made from the flowers of the lavender shrub to such an extent that while supplies are dwindling, countries in the old world, like Bulgaria, Hungary, and Greece, try to keep up with the production of the most sought after true essential Lavender oil because the traditional lavender fields of Provence in France have been severely damaged in recent years by insect-borne bacteria. Since I am happy to grow in rocky and sunny terrains, I am welcomed in Australia and Somalia recently.

While my favourite special sister *Lavandula angustifolia* is blooming only after some years, in higher more difficult to grow regions, my purple flowers are ready to be harvested for oil production right in my first year. Me, the sturdy soothing mule from the lavender family.

Ildiko Meny, M.D., M.P.H.
Public Health Consultant
mesmotsdanslanuit.blogspot.com

MENTHE PIPERITA

Peppermint

*Captured by the presence of a young woman standing
in the middle of a peppermint field in rural Wisconsin,
our gaze rests on the potted plant poised on her palm.
Her right fingers squeeze the edge of the pot close to her
shoulder blade. This plant is of an unknown kind. Next
to this photographic image is a drawing of a plant in
carbon on found glass, which leans at an angle against
the wall. Holding these two images is the presence
of the plant in the photograph on a wooden plinth.*

Introducing *Menthe piperita* L. (peppermint), the offspring of
Mentha spicata (spearmint) and *Mentha aquatic* (watermint).
This naturally occurring hybrid spreads rhizomatically,
having earned him/her the humanized categorization
of a 'weed.' Creeping just above soil level, it sends out a
series of runners or stolons from which new plants are
continuously propagated, emanating nearby to the parent plant.
I became interested in peppermint, through a little-known gardening
history in America with parallels in England, known as 'Gamma
Gardening.' This gardening technique was established in various
scientific labs throughout the world to test the effects of irradiation
on plant life. Success with unsuspecting beneficial mutations
encouraged wider use. Searching through an online database,
I discovered two strains of peppermint: 'Todd's' Mitcham and
'Murray' Mitcham. They had been irradiated and patented in
1955. Both of these stemmed from the original parent plant

'Mitcham,' a very old variety of peppermint whose origins are English. The names 'Murray' and 'Todd' reveal a diverse lineage of scientific and agricultural practice in the United States. Merritt J. Murray was a scientist who worked at the Brookhaven Lab, where the first 'gamma garden' was established, and a researcher for the A.M. Todd Company. 'Todd' revealed one of the key families in the States involved in peppermint production, particularly in Michigan: as a young man of 23, Alfred Todd travelled to Europe where he found himself in Mitcham, a small town near London. Walking through fields of mint, his nose noted the subtle difference from the plants he grew back home on a small patch of land. A few years later (1890s), he shipped in bulk peppermint rootstocks, cultivating a thriving peppermint industry in America at one time.

Walking through the William and Lynda Steere Herbarium of the New York Botanical Garden, I arrived on the fourth floor, guided by the archivist to rows of large standing metal presses catalogued to the *Lamiaceae Family [285]*. Between the pages of the first folder, I found various samples of 'Todd's' Mitcham and 'Murray' Mitcham grown at A. M. Company, Kalamazoo, Michigan, dated 1955. Given this rare opportunity to draw directly from these plant samples, their earthy smells lingered as I traced the plants edges. Their essence resonates in this temporary installation, and their mutable traces are re-imagined over time through the stresses and tensions that prevail around human-centred ethics and the assumptions that underpin scientific progresses. Yet, a simple drawing materialized.

Christine Mackey

Christine Mackey, *Looking for Todd*, 2018 →
Peppermint farm, Wisconsin, USA

M I M O S A PUDICA

Touch-me-not

The *pudica* is a pose. She is a figurative, silent representation of an interior state. As a studied attitude, the *pudica* is seen through the spectator's gaze, seeing herself. She is formed by the spectator, oriented toward violation, protesting the touch that has already taken place. The more theatrically the *pudica* refuses herself to us, the more we access her inaccessibility, and in doing so we fulfil her function.

This refusal—and the suspension of the moment before it is swept aside—constitutes the eroticism of the *pudica* in which the distance between the one who desires and the object of that desire is held infinitely just shy of contact. (In that eternity, the promise that contact will be made is simultaneously legible.) The *pudica*, too, is shy. All her blood is on her skin; she is utterly revealed, transparent and fully lit, her genitals outside of the body. Her attempt to cover herself falls short, too. A set of three dots measure the space between her palm and her breast, between her digits and vulva—an ellipsis of air.

Mimosa pudica is an inconspicuous plant, noteworthy first for its motion (which is a perceptible, reproachful drawing-in of the leaves at the touch of a finger), then for the reading of a particular affect in that motion (the plant's name includes "shy," "sensitive," "humble," "shameful," "touch me not," and "noli me tangere"). We interpret the plant's responsiveness as a form of protest, as a rejection of contact, in the moment when we also acknowledge our desire for contact with it *in order for it to manifest its response.* We want to see ourselves recognized; we want to know that we are heard, seen, and felt, that we have made an impact, even if our touch can only be damaging. In describing the plant as ashamed, sensitive, shy, and then touching it anyway, we assign it pain. The motion of the *Mimosa pudica* is not only its shrinking and closure, however. In South America, where it is native, it is called *morí-viví* for its pantomime of dying and returning to life. The plant, therefore, has two expressions—that of registering contact and that of resurrection. Perhaps it is also important that a pose is a pause, that it is a moment not in time but extracted from it, elongated, and stylized. The fast-moving plant, its vegetal hum transposed into an animal pitch, elasticizes time, rotates it, returns it onto itself. The figure of time for *Mimosa pudica* is not the arrow but the ellipse, the rotating triangle defined by three points: the desirer, the desired, and the infinite reaching that comes between them.

> Note: This text accompanies the installation Noli Me Tangere (2013), in which a Mimosa pudica is held open in an otherwise dark room by the light from a video projection of a hand drawing an ellipse. Pudica, a Latin word meaning shy or ashamed, is an art historical pose depicting a nude female figure covering herself suggestively.

Caroline Carlsmith

Caroline Carlsmith, *Noli Me Tangere*, 2013
Mimosa pudica, projector, video, boxes, soil
(same on p.56)

MORS ONTOLOGICA

Death of the Spirit

Timothy Morton, in *Dark Ecology*, unpacks the serious play of Substance D, "the drug of meta,"[1] featured in Philip K. Dick's 1977 novel, *A Scanner Darkly*. Substance D is synthesized from the fictional little blue flower, *Mors ontologica*—the pseudo-Latinate translating to 'ontological death' or, idiomatically, "Death of the spirit. The identity. The essential nature."[2] In Dick's novel, readers trace drug user and dealer Bob Arctor, alternately (un)known as Fred, an undercover agent informing on himself, and the burnt out "poisoned husk" Bruce, a casualty of Substance D, through his disenfranchisement into corporate slavery at the oppressive mechanism of industrial agriculture. Participating in the genre of popular 20th-century plant-horror, *A Scanner Darkly* ports into a grander posthuman, ecocentric narrative—that of the interspecial, sociosexual expression of plant desire in the gut of the human mind; of the little blue flower, *Mors ontologica* and its complex interrelationship with human social structures and psychologies. These "lovely little blue flowers"[3]—revealed at the novel's twist to be the obscure organic origin of Substance D, the Slow Death—demonstrate the uncanny double bind of plant blindness and elucidate the insidious power of plants to subvert Enlightenment preconceptions of individuated human autonomy. In the final, climactic scene of the novel, in which now-Bruce sees fields of furtive flowers sown among rows of corn, the

Executive Director of the New-Path rehabilitation facility, in an act of vegetal prestidigitation, vanishes whole fields of *Mors ontologica* from before the New-Path labourer's eyes. In a curious reveal, the Director explains, "No, you simply can't see them. That's a philosophical problem you wouldn't comprehend. Epistemology—the theory of knowledge."[4]

A Scanner Darkly, with its claustrophobic, paranoiac storyworld, in this flowers-all-the-way-down reading, demonstrates three various vegetal *épistémès* embodied in one corpse, three disparate subjects, with three disparate subjectivities, interpellated by the little blue flower: the consumer, who knows not the plant but the mediated product as an end in itself; the retailer, who knows not the plant but the product as a medium to profit; and the producer, who knows the plant and the product, too, but is disavowed their resemblance. External to, yet indivisible from this tripartite subjectivity is the capitalist who knows the plant, the product, and their process and polices the knowledge of the relationship there between.

In this way, *Mors ontologica* demonstrates the vegetal foundation of the very constitution of the human subject, however divided—the plant paradoxically effecting individuated subjectivity whilst occluding any self-knowledge thereof. Its psychosexual expression peoples novel constellations of being while burying its radical origins under its amnesiac living death: a condition that Morton identifies, citing Giorgio Agamben, as *bare life*—"Pure survival without quality, based on fear, generating people who can't tell whether or not they are people working on objects they can't tell are objects."[5]

Yet, such a pretty little flower poses a challenge, both ontological and epistemological, to contemporary understandings of human and plant natures—a challenge (perhaps to ontology and epistemology themselves) to re-cognize novel conceptions of life, however nude or draped in the trappings of the age, as constellated *among* and *through*, becomings of the various agencies of the beings of the world: an ecognostic re-thinking of the misconception of human independence *from* and mastery *over* one's self and a more-than-human world.

David M. J. Carruthers
Ph.D. Candidate
Department of English, Queen's University, Canada

OPLOPANAX HORRIDUS

Hoolhghulh

oh—

not like knowledge—oh
                    against th—
                an oth—

        (caught up—

        (the abrasion of—

        (skin against skin thinks—

        (trans—

                loss)

an old
growth
pine/spruce/balsam
context a fine spray of light

the register of the thing
ness, and then some
thing else
there
leaves, leaving / outside
conscious of—

oh, this is where
the the

begins   (ow

<cultural contact zone>   an opening

an opened up thigh
an open discourse (vascular xylem phloem
an open poem pouring

(ow what the
who

when Hoolhghulh spoke
my relations changed

with respect to being

a recumbent root growth
across—a time signature—tendril
senses—fluid rhythms—soil air

when I stopped 'reading' the land
when Hoolhghulh spoke

a vibration on the skin
of my hand held
over a leaf face

affinity a fierce
loyalty—a companion
plant sharing a space

Hoolhghulh spoke.

on the personhood of the plant

they reside and breathe in time,
discuss the reasons to move,
feel tiny changes in air and hope

this one

tells me to pay attention
a charge of energy from their leaf to my hand
my eyes

drawn

they look at me

this one smells my air—molecules circulate,
other senses and there is a requickening
like over long distances or over alien tongues, phloem

they have the ability to not be seen—the song
or offering, the choice is their green divination,
a gesture to
this one (heavy-footed but alert

the contact

the choice of pronoun
versus the presence in being

the full sentence across a page or
across a season I

am piteously incapable, incapable
when human metaphors fail

what is left

fibre returned to soil

(breathe)

what measure and by whom
does your suffering occur?

or, what language sends its
alchemical signals out, the sense
in a tip of rootlet, the genius
in that relationship with the fir

you stand before this creature, me,
asserting territory, asking for
protocol, wary of my intent

(you have sent a quick warning
to others of my presence)

the contact

in the fragility of this fleeting open
moment—a finger a tongue an eye—
living flows and
old age approaches
you think about your young

the contact
where the exchange in different
languages: electrochemical not inglish
and back again a greeting
made specific in this moment this
place of moss and sword fern fanned
by this particular breeze and scents

how to suspend the humanity in that
moment?

you share water

the contact

this zone now a place of light sandy loam inland river-formed bank second
growth (colonial imprint and response)
zones of our ideas of each other
(the shape of the plant's idea)

zone of thought turned inward the human refrain turned off—
a dissembling of the historical body standing in the face
of the leaf's poise, the scent's call

zoned out from the language, I hum a response (hand held out
over a leaf, an electrical greeting across cultures

hello

Dr. Robert Budde
Professor, University of Northern British Columbia

ORYZASATIVA

Rice

Floor art practices in India are over 2000 years old and have been documented since AD 50[1]. Using a white powder made from rice grains, floor art practitioners create threshold and courtyard designs. These can be seen across various regions in India and are referred to by different names, such as *Kolam*, *Rangoli*, and *Muggu*. Created daily, these simple line drawings become increasingly elaborate and intricate during celebratory occasions. Ritual art practices using rice continue into the present, especially in the south where the drawings are specifically referred to as *Kolam*.

The rice grains are used in powdered form. This endows them with the natural property of ephemerality, an attribute that aligns with the idea of impermanence in Hindu philosophy[2]. The edibility of rice makes the *Kolam* a form of food offering to ants, birds, and small insects, an act that lends itself to the Hindu religious idea of *ana-dana*, or sharing of food. There is another belief that the Hindu goddess Lakshmi had her dwellings in grains of rice, thereby making its powder auspicious. The enactment of the *Kolam* thus becomes a ritual that brings the *Kolam-maker* closer to the goddess[3]. The whiteness of rice symbolizes purity, and the entrance of the house is duly sanctified by the presence of rice on its threshold[4].

Kolams made with rice powder are created using dots and lines in geometric and curvilinear forms. These forms represent images, such as conch shells, flowers, leaves, the sun, and the moon. They are created daily, generally by the women of the house, and range from simple diagrammatic representations to more complex decorative patterns that are often adorned with real flowers[5]. Rice powder as a botanical material is encountered in this way across many residential streets in south India, through a highly aesthetic ritual art practice. Botanical art is specifically concerned with the study of plants through art practices. The field was established to serve the needs of botany during colonial expansion where plant illustrations assisted with plant identification. Scientific accuracy and anatomical precision are characteristic features of botanical art practice, and this convention has dominated the visual expressions of plants through herbals and botanical books[6]. It, therefore, does not consider the aesthetic and indigenous uses of plant forms themselves, as in the ritual practice of *Kolam*. Rice as a material of artistic expression, used for its sacred and symbolic properties, circulates an alternate aesthetic engagement—one that is rooted in myth and imagination in contrast to the desire to capture and control.

Touch, pour, and draw with the plant...

Let the powder slip between your thumb and fingers.

Geetanjali Sachdev
Art and design pedagogue
Srishti Institute of Art, Design and Technology

Kolam and Rice, 2018 →
Ann Rosenthal, Independent Artist and Educator,
Locus Art Studio

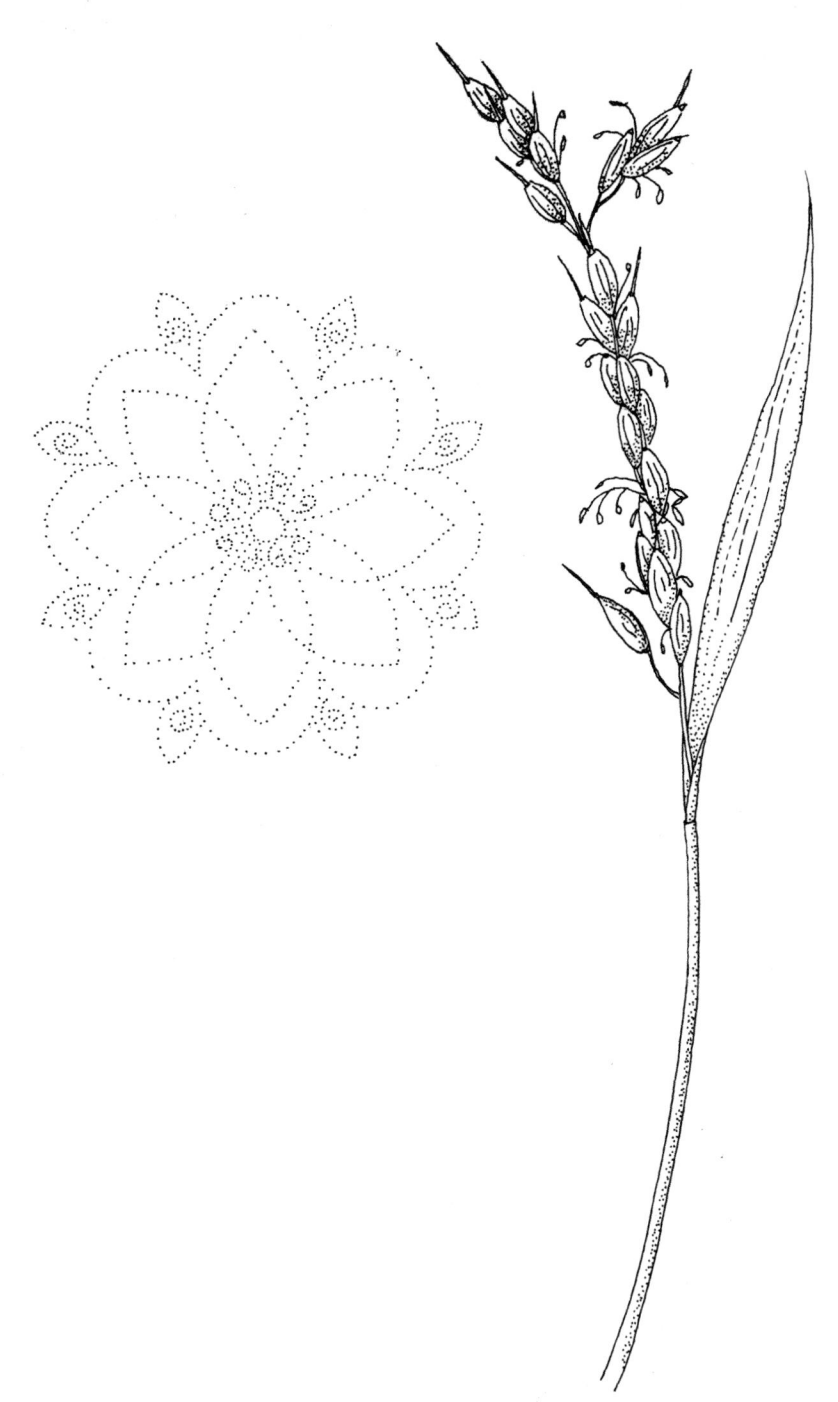

PERSICARIA

Lady's Thumb

Red bamboo-like silk stalks. And strong leaves to either side
    that yellow as I reach openly to the sky. But I can be bigger
    than you imagine. Not just that red marking, the rubbing
    that everyone looks for. And yes, you can press your finger,
    your body into me, and I change like a brilliant mood ring.
Atop a pink crown of tiny fruits. KWEEN they call me. But not A QUEEN.
    And yes, QUEEN I will certainly take sometimes. But not all the time.
You are always so concerned with gender just because my
    seeds are pink. But they are also fuchsia and red and
    white and dark. I am not a lady. OR YOUR LADY. More
    than a thumb. Our stems unfold in sections, gathering
    strength. With leaves that can bend in the wind, with
    water. And, if you look close, tiny hairs under each appendage.
In this garden, we are of course dwarfed by the mother patch,
    the mugwort that surrounds. She is a careful organizer,
    though, who provides shade and fungal transfer, a below
    ground for many organisms that call this place their home.
    We are a resilient KWEEN, even in the part of the garden
    that you call the desert—we thrive and find space just as our
    knotweed sisters have done for centuries. Alone together.
    In patchiness and solitary growth. And I know you use us
    for wounds, for rheumatic pain, to ease stomach aches.
    To eat of course; yes, the Vietnamese call us coriander indeed.
But what now of that surrounds? We can hear planes passing overhead,
    the rumble of the nearby subway, the beat of neighbourhood
    gatherings. Thump and bass. Our leaves are not as green as

they once were, yellowed from heat and compacted soil filled with so much. Why do you leave all these metal bits and seal the ground with rocks and glue and tar and plastic? Still, our seeds find cracks, and we are crowned again. And where did we come from? Oh honey, this story is long. Sometimes the wind and water flows and birds who adore our pink fruits. But we come from the Eastern shores, from a place now forgotten here in this city place. And what of movement? My darlings. You must walk today like a KWEEN. Not the kind you are thinking of but the QUEEN down in your bones. Begin by standing still and closing your eyes. Feel your shoulders; let them drop, let them rest. Now roll your head clockwise, and then reverse. Slowly. Remember to breath. And think of what you need to let go of, think of what you want to be. Hold on to this image, hold on to this moment. Straighten your whole body. Look up and then forward. And now strut. Walk with your chest out, a crown on your head. You are a Pink KWEEN indeed.

Christopher Lee Kennedy
EPA Agent, Environmental Performance Agency

PICEA GLAUCA

Mina'ig
Highland Spruce

Anishinaabe obagidinaan Asemaan waabanong inake'ii ini Mina'igoon ji-bizindaagod aaniin inake'ii ezhi-maanendang awe Anishinaabe: An Anishinaabe offers tobacco on the east side of Mina'ig so that s/he will be listened to as s/he explains feelings of sadness. *Aapiji iidog gizhewaadizi awe Mina'ig highland spruce izhinikaazo gii-wemitigoozhiimong.* S/he is very kind, Mina'ig, this one who is called *Highland Spruce* in English. *Miich owe gaa-izhi-gikino'amaagowaan owe, owe gegoo izhi-maanendamaan, gegoo izhi-goopadendamaan gakina gegoo ezhi-zhiingendamaan.* This is the way I was taught: when something makes me feel sad, when I feel like things are worthless, when I just hate everything. *Naa mii imaa nake'ii ezhi-asag awe waabanong inake'ii awe mitig, mina'ig ezhinikaazod, ji-bizindawid awe Manidoo-mitig weweni ji-ani-ezhi-ayaayaan owe ishkwaa-maanendamaan.* That's when I put down Asemaa on the east side of this tree, the one who is called Mina'ig, so that this tree spirit will listen to me, so that I will be good, so that I will stop being sad. *Naa mii owe gaa-izhi-gikinoo'amawid bezhig niij-anishinaabe. Mii owe ingikinoo'amaagoowin.* That is how my Anishinaabe friend taught me. This is my teaching.[1]
My mother, Mary Siisip Geniusz, taught me that every being on this planet has a spiritual and a physical purpose. She first received these teachings from Keewaydinoquay, an Ojibwe medicine woman from Lower Michigan. In *Anishinaabe-inendaasowin,*

Ojibwe philosophy, those beings include *gaa-nitaawigiwaad omaa akiing*, those who grow here on earth, the plants and trees. According to Anishinaabe-inendaasowin, not only are plants and trees living, sentient beings, they are also our elder brothers and sisters because they were created before humans ever existed. Plants and trees are closer to *Gichi-manidoo*, the Creator, and they are far more powerful and have more wisdom than humans. Whenever we want to ask assistance from another being, we must make an offering of *Asemaa* (tobacco[2]) or *Kinnikinnick* (a plant mixture) to that being, explain what we want, and then wait for a reply. This is very important when asking for spiritual help from plants and trees. If someone uses part of a plant or tree without first asking for that being's permission, that person might only receive physical, not spiritual, assistance. When healing, not receiving spiritual assistance means that someone is not fully healed.[3]

The teaching presented at the beginning of this article, told by Waasebines, Kenneth Johnson, Sr., from Seine River First Nation, tells us about a spiritual gift that Mina'ig has to share with *Anishinaabeg*, Ojibwe people, Indigenous people, or humans. When talking about Mina'ig in English, Waasebines often said, "Mina'ig is a good listener." As we search for ways to reconcile negative colonization forces of the past that continue to shape our future, we, as Anishinaabeg, need to remember that we are not alone in our battles: Mina'ig and our other elder brothers and sisters are right here with us. If we treat these beings with respect by preserving their living spaces, if we bring them offerings and explain to them the problems we face, they may choose to work with us.

Wendy Makoons Geniusz, with Waasebines, Ken Johnson, Sr.
Associate Professor, Languages
University of Wisconsin Eau Claire

Annmarie Geniusz, *Mina'ig*, 2018

Plantain leaf as bandage for abbrasion

PLANTAGO MAJOR

Plantain

The humble plantain leaf is a testament to the complicated histories and profound healing one can find in the most unassuming places. Eurasian in origin, Broadleaf Plantain has been a treasured part of many healing modalities since ancient times. Plantain appears in Traditional Persian Medicine,[1] and studies of peat bogs show that it existed in England before recorded history. It appears in many great artistic and literary works including Shakespeare and Durer. Plantain seeds found in the stomachs of preserved Danish 'bog people' from the 3rd and 5th centuries C.E. show that our companionship with Plantain stretches back millennia.[2] The 10th century Saxon *Lacnunga* cites plantain as one of nine sacred plants (*see inset*), focusing on its indestructibility, which was a portend of its move overseas many years later.[3]

In Europe, plantain primarily served as healer, but in North America, it has a dual identity. Broadleaf Plantain moved to the continent with European colonists, becoming so prolific that it was referred to as "White Man's Footstep".[4] Like the Europeans who marched across the land, laying claim on everything they found, so too did plantain move across the continent, laying down roots in soil already occupied by others, altering the land wherever it took hold.

However, plantain's work on the landscape was decidedly subtler than the colonists who transported it. A short, stubby plant with graceful stalks and big, round leaves, plantain is memorable for

And you, Waybroad [Plantain], mother of herbs,
open from the east, mighty within.
Over you chariots creaked, over you queens rode, over you brides cried out, over you bulls snorted.
All this you withstood, and confounded.
So you withstand poison and flying venom, and the foe who goes through the land.

how immemorable it is: Nearly everyone here has encountered it, but only those who know it consciously register its presence. Broadleaf Plantain's modest appearance belies its incredible healing power. Plantain is a food and a healer of everything from bug bites to sore throats to constipation. It can also be understood as a bridge between medicinal traditions. It has been incorporated into some Indigenous medicinal traditions and into blended folk medicine traditions (e.g. lower Appalachians). The knowledge of how to use this and other healing plants was shared between Indigenous and European communities, each contributing knowledge to assist the other. Robin Wall Kimmerer offers an apt description for this symbolism, comparing human colonial behaviors with those of the plant world:

> *Foreign invaders like loosestrife, kudzu*, and cheat grass have the colonizing habit of taking over others' homes and growing without regard to limits. But Plantain is not like that. Its strategy was to be useful, to fit into small places, to coexist with others around the dooryard, to heal wounds…Maybe the task assigned to Second Man is to unlearn the model of kudzu and follow the teachings of White Man's Footstep.*[5]

Julia Skinner
Founder & Director, Root Kitchen
@rootkitchens

*See entry, *Pueraria Montana var. lobate—Kudzu*, p.93

PLATANUS × ACERIFOLIA

London Plane

Berkeley Square planes

I don't believe nightingales sang here
and I need something solid in this
sticky shade,

not full throated ease and plaintive anthems
fading far away, dissolving
and making me quite forget.

Give me mottled trunks, flaking only to save us
from soot, and branches baubled
in celebration;

our rescue from the smog, they're banishers
of the pea souper,
re-buffers of lung burn and cough,

visions more dappled and magnificent
than any waking dream.

Kate Noakes, M.Phil.
Poet
Latest collection: The Filthy Quiet, Parthian, 2019
boomslangpoetry.blogspot.com

Polianthes tuberosa

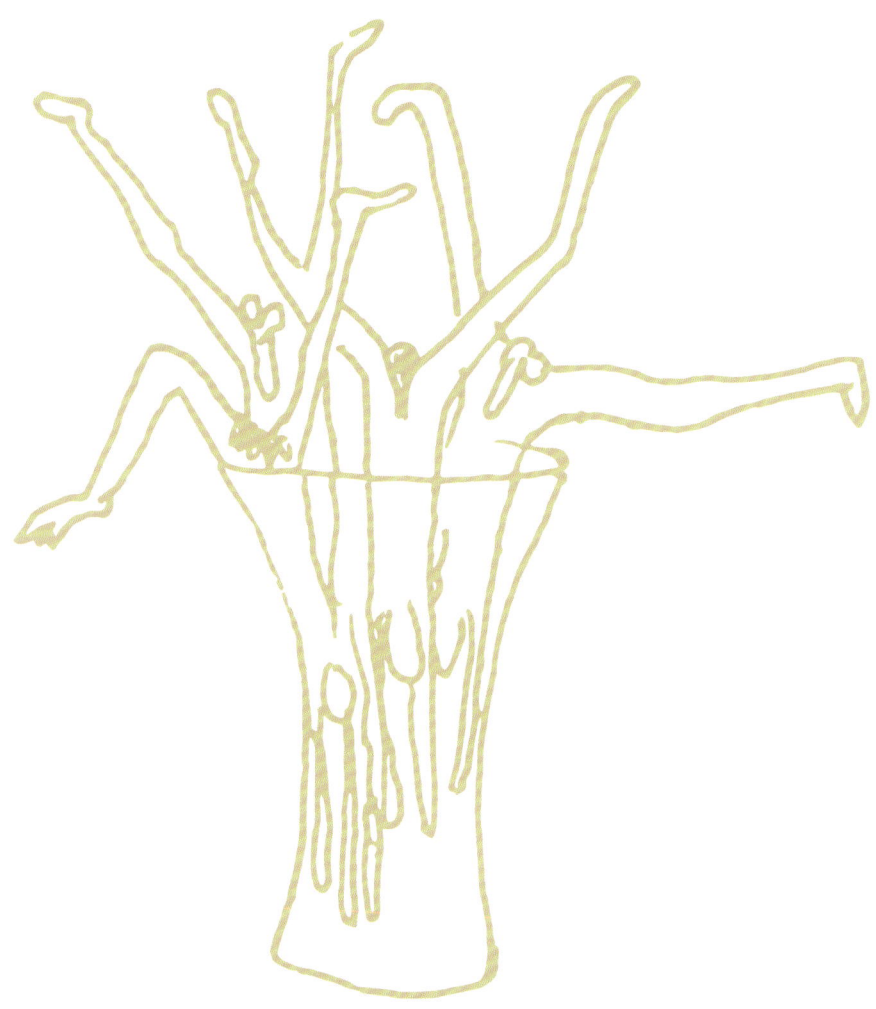

Unknown, *Flowers are the genitals of plants*

P O L I A N T H E S TUBEROSA

Tuberose

The story has been reproduced with enough frequency that it has become virtually untraceable: Victorian girls were forbidden from smelling the bloom of the tuberose for fear they would be overcome by spontaneous orgasm. *Spontaneous*, as in spontaneous generation, meaning outside the domain of reproductive sexuality. In other words, a whiff of tuberose initiates a deviant sequence of effects that might end in new sexual allegiances outside the human domain. In his 1905 study of the role of the senses in evolutionary attraction, *Sexual Selection in Man*, Havelock Ellis quotes the 19th century physiologist Paolo Mantegazza on this sequence: "Make the chastest woman smell the flowers she likes best, and she will close her eyes, breathe deeply, and, if very sensitive, tremble all over, presenting an intimate picture which otherwise she never shows, except perhaps to her lover." Floral pleasure here is framed as akin to masturbation's solitary pleasure. A deeper suspicion though is disclosed in the description's final clause— "except perhaps to her lover"—the sense that this pleasure is anything but solitary, that the flower is itself a strange new lover. The waxy flower of *Polianthes tuberosa* is indeed heady and "penetrating," in the words of a woman quoted by Ellis elsewhere in his study, but the tuberose is only the emblem of other lurid florals: contemporary accounts note as well the whetting and moistening effects of gardenia, lily of the valley, orange blossom, and other white flowers. These are all flowers whose heavy fragrances suggest a line of seduction away from human-based feeling toward a range of sensations of unforeseen intensity of which the flower's bloom might only be the first. I want to float this story from the position of pleasure rather than prohibition—too easy and too familiar to extrapolate a microhistory of limitations on women's pleasure—and for this pleasure to float, the interface of flower and lover operates as fantasy. The orgasm provoked by the inhalation of a flower then is not an anxiety of propriety but an aspiration. Wouldn't we like to abandon the project of human sexuality? Yes, we would! We need our tuberose then, our bloom that shows us through erotic feint the tenuousness of our attachment to human stimulation. Wouldn't we like to show the blossom our hidden face? Yes, we would.

Willy Smart

I consulted the Mysteries the other day and access was not granted. The image is of a termite eating the edge of a book.

Access was not denied wholesale, but with deep respect, access was not granted at this time. The notion activated here is that knowledge grounded in context and rooted in relationships of mutual interest is the kind of knowledge we want to cultivate.

Knowledge is contextual or it risks universalism. Who will pick the book up? Where are they located? What is the nature of their relationship with us? Divination reminded Jamie that what we have learned about surrogates to real relationships are not useful. We urge those who read this to humble themselves before teachers, human and plant.

A species exists together with every other living thing in its ecosystem. The whole messy thing is a plant, alive, dying, fucking and giving itself to heal us, feed us or poisoning us. The plants that work together in the spell that we are crafting constitute their own sensitive ecosystem. They are greater than the sum of their constituent parts, their names.

IRRUPTION #2

Put this book down. Stop reading.

Remember your body. Have you been sitting for a while? Are you hungry? Are your breaths deep or shallow? Close your eyes. Put on appropriate clothing and move outside. Take this book with you.

You're not going to tell anyone where, why, or when.

This ritual is as old as time. Don't worry about what people have called it. Defy the need to explain what it is that you're choosing to do with your time. Allow this to not have a name. This is not 'going for a walk', and this is not 'gardening' or 'bird-watching.' Put on appropriate clothing and move outside.

You're reading this book because you are drawn to plants. That's a great first step to what we believe is needed in answering the question we're looking to help have answered.

We would do very well to listen to where our interests lead us. We should feel the soft drawing power. The drift has begun. When you get there, you'll know. Maybe it takes you 10 minutes, all afternoon, an entire lifetime.

You're looking for a spot directly on the ground, the foot of a tree, a rock, soft moss. This is a place where you can advocate for yourself, and speak to the world more effectively. Sometimes these are beautiful places. You may find yourself drawn to stillness as if 'zoning out', like a passenger in a moving vehicle. You may find a bit of soreness, similar to when one palpates the skin, searching for the precise acupuncture point: the meridians running up and down the human body mirroring the ley lines that span the globe. These are the soft places, the points you can stimulate change. Jamie follows the signs of synchronicity searching for these points of connection in his city soft spots in his city. For him, déjà vu and the sensations that arise on the surface of his skin are his cues.

When you can't identify a plant in a survival situation, there are a series of steps to determine if the plant has the ability to kill you. You approach it gradually coming in greater contact with it, first visually, then with your olfactory senses, with touch, and carefully, with time between each attempt, gradually introducing tiny amounts into your mouth before swallowing it. There's no rushing this process. Introduce yourself to the place where you find yourself. Own your interest in going deeper with plants—what do you want?

Tear this page out of the book. Put it on the ground in front of you. Step onto it. Look around. Are you already familiar with any of the plants you see? What are their names? Are any of the plants you see used traditionally as food? Are any of the plants you see endangered? Invasive? Take a moment. When you're done, place the page in the ground.

Jamie Ross
Contemporary artist, diviner and witch

Gesig Isaac
Multidisciplinary Mi'gmaq artist from Listuguj First Nation

POPULUS
DELTOIDES

Cottonwood

Could a cottonwood be a conversation partner?

They are certainly more than fluent speakers. Their very
leaves sing, flowing water in the Kansas wind currents.
I started to grow up underneath two cottonwoods. They still stand at
the bottom of a slight hill outside my childhood school building
where the water would pool. Every day when the recess bell
would ring, we would run and sometimes roll down to play in
the flickering sunlight beneath their leaves. In late spring we
would gather the "cotton" of the catkins where we could reach
or try to pick it out of the air as it floated by, and in fall we would
gather the leaves with their yellow scent bright against the blue sky.
I have never planted a cottonwood, only played with them. I would
not seek them out in a prairie lightning storm. But I know
they spoke survival, signing a source of water, to people like
my ancestors, immigrants coming over to the plains. And
I've learned how they spoke, and continue to speak, sacred
power to Indigenous peoples, relatives of theirs rising on the plains.
In the early twentieth century, the Lakota elder Hehaka Sapa, Black
Elk, explained that cottonwoods offer shelter and support.
They are trees of mystery. In the music of leaves in the breeze
"you can hear the voice of the cottonwood tree," he said.[1] The
rustling, shielding, flowering, singing cottonwood stands at the
center of the Sun Dance and the Great Vision that Black Elk speaks.

Umónhonti—the real Omaha, the Venerable Man—is the cottonwood
Sacred Pole. He has a relation to all Omaha people that is
spiritual and storied. In one version of the Omaha people's
Sacred Legend, recounted by Robin Ridington and In'aska
Dennis Hastings in their book *Blessing for a Long Time*, "He
stands at a place where the four directions, the night sky, and
the earth come together to a single point. He is the one who
makes the number seven. He is an individual who stands
for all the people, a singularity who stands for an entirety."[2]
He was repatriated to the Omaha in the late twentieth century.
Black Elk said the cottonwood taught the Lakota how to
make tipis. The ethnobotanist Melvin Gilmore took
a photograph of the creation of Omaha children: a
húthuga, a camp circle, made of cottonwood leaf homes.

Cottonwoods grow up quickly. My own learning seems slow in comparison.

I still stand listening to these teachers of fluency, to their celebration
of being at home. Will I ever converse with a cottonwood?

Let me be content with the useful experience of immersion.

Aubrey Streit Krug
Director of Ecosphere Studies
The Land Institute

PUERARIA MONTANA VAR. LOBATE

Kudzu

The concept of The Other refers to hierarchical social order established and maintained through the creation and representation of simple binary *us* and *them* oppositions.[1] In recent years, a worldwide right-wing backlash over an ongoing global refugee crisis has reacquainted many with the image of the *alien* outsider who, we are told, is *invading* and *taking over our way of life*.[2] In the socio-political realm, we may recognize this rhetoric and understand its historical manifestation into bigotry and violence the world over. One might ask then, how is it that terms such as *invasive, native, alien, naturalized* etc. are deemed necessarily ripe with ideology and prejudice in a sociological attempt to understand human migration but at the same time form the fundamental vocabulary of biologists in describing the migration of all *other* species in the natural world?[3] To this extent, the very terms of conservation biology, that species migration should be understood in terms of foreign invasions of otherwise pristine and balanced *native* habitats is extremely problematic. In both cases, there is often a failure to account for larger forces at work in which migrations play out as a symptom rather than the condition. It is easier for the U.S. government to demonize immigrants from the middle east and south of its border, for instance, than it is for it to look critically at its foreign and economic policy. Likewise, it is easier to demonize migrating plants and animals as *invasives* destroying

← *Kudzu graveyard*

our *native* habitats than to look at causes such as climate change, overdevelopment, and toxification of the biosphere.[4] We owe this language to the cold war era invention of invasion biology, which helped to codify a conservative, nationalistic, and militaristic language to describe the rapidly changing ecosystems of the 20th century. As science grew in its awareness of the war colonization and industrialization had waged against the earth's life support systems, so grew the reactionary response of conservation science to freeze those systems in a constant state, as though change itself were the enemy. The colonizer is recast as the saviour, and yet the colonizer's desire for control continues.

Fortunately, a new generation of ecologists, documented in such books as Fred Pierce's *The New Wild: Why Invasive Species will be Nature's Salvation* and David Holmgren's, *Beyond the War on Invasive Species: A Permaculture Approach to Ecosystem Restoration*, are asking if many of those perceived alien invaders, with their tenacity, rapid biomass production, and ability to grow in the most inhospitable conditions, might be part of the answer to some of our most pressing ecological problems rather than their cause. The question is, will we look to the value of the *aliens* in our midst and see what they might have to offer for our *new wild* spaces, or will we continue to wage war aimed at conserving or recreating a natural history that may be ill-equipped to deal with a rapidly changing planet?

Known as the "the vine that ate the south," Kudzu is the prototypical "invasive species" in the United States, and as such it provides a good and challenging starting place to rethink the "native" versus "alien invasive" paradigm. Kudzu was introduced to the Americas from Asia in the 19th century and used as a high protein cattle, goat, and sheep feed, as well as to prevent soil erosion. During the 1930s, the U.S. government paid farmers to grow more than 1.2 million acres of Kudzu to counteract the effects of erosion, and by the end of World War II, Kudzu covered an estimated 3 million acres of land and was heralded as a key to US recovery from the "dustbowl."[5] The postwar "green revolution," the expansion of farmland, residential development, and the ceaseless building of roads in the eastern U.S., created the perfect habitat for Kudzu, which thrives in forest edge habitats with depleted and poor soils. As Kudzu's benefits became obsolete to human interests and as humans cut more forests, creating more edge habitat, Kudzu began to be seen as an invader rather than a savior. By the 1970s, Kudzu's biological success was met with a human downgrade in status to "noxious weed" and became the target of extermination campaigns armed with huge quantities of toxic herbicides, which seemed to have virtually no effect on the plant.[5] However, despite being mythologized as "the plant that ate the south," a recent U.S. Forest Service study found the plant to cover only 227,000 acres (about 1/6 the size of Alabama) of land rather than the millions usually quoted, and far less than in Kudzu's heyday of the '30s and '40s when it was heralded as a savior.5 Meanwhile, more than 97,000,000 acres are devoted to corn in the U.S. (about the size of California), virtually all genetically modified and virtually all grown in monocrops where any other living thing is considered an invading organism. And yet, corn is not an invasive plant.[6] This fact exposes the ideological value system applied to the supposedly objective science of studying ecosystems and the ill-conceived separation of agriculture from "natural ecosystems" in our culture versus nature paradigm. Considering Kudzu's enormous use-value (perhaps on par with corn), the "problem" of Kudzu can be rephrased as one less concerned with invasion and more concerned with underutilization.

Mark Cooley and Elizabeth Hall
SporaStudios, George Mason University

Punica granatum

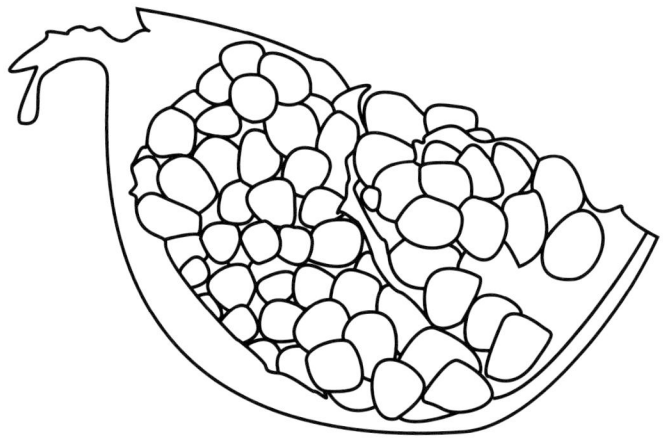

Lili Huston-Herterich, *Pomegranate*, 2018
Vector Illustration

PUNICA GRANATUM

Pomegranate

Throughout the historical world, a variety of methods for contraception were known to ancient peoples. Manual interventions with the body—like coitus interruptus, breathing and sneezing techniques, male castration, and condoms—were employed alongside medicinal ingestion and application of materials from the natural world, including copper ore, mercury, animal dung, and the leaves, roots, flowers, fruits, and seeds of various plants. Pomegranate was viewed as a symbol of love, sexuality, and fertility with its ruby-red conglomerations of seeds protected within a womb-like skin. It is referenced in texts and mythologies in Ancient Greek, Hebrew, Christian, Buddhist, and Islamic traditions. In West Asia, it represented the goddess of love, but in Greece, it stood for sterility. The first medical records indicate that it was employed in treatment for tapeworms in 1500 BCE. A later 11th century CE Arabic text by Avicenna lists pomegranate as a post-coital contraceptive to be administered as a suppository.[1] Soranus of Ephesus, an Ancient Greek physician active between the 1st and 2nd centuries CE, provides this recipe for the use of pomegranate as a preventative method: grind the skin of a fresh pomegranate, add water, and apply to the vagina. Alternative recipes involve oak galls, rose oil, and gum.[2] Experiments in the 1930s by Adolph Butenandt and H. Jacobi confirmed the fruit contains a human sex hormone that reduces fertility in women.[3]

Stella Brown
Artist

R A P H I A

Raffia

Raffia fibers are derived from the Raphia palm trees of the tropical
regions of Africa—particularly Madagascar. Raffia fibers are
produced by stripping the flexible membranes of each palm
frond from the tough inner layer. The palm skins are then cut
or stripped into widths appropriate for their use and dried in the sun.
Throughout the world, raffia is used as twine and rope, as well as
in the production of hats, bags, placemats, and textiles. While
raffia has become a prominent global export from the jungles of
Madagascar, raffia plays a prominent part in the culture, beliefs,
and rituals of many indigenous societies in Africa and beyond.
Between 1995 and 2003, Fern Shaffer and Othello Andersen enacted
nine rituals throughout North America to transmit healing
energies to landscapes which have, or are experiencing,
ecological trauma. Within a costume of raffia and strips of
canvas adorned with screws and bolts, Fern intuitively danced,
sang, chanted, and prayed for the Earth's healing. Just as many
humans before her, the costume of raffia allowed Fern to escape
the mundane, relinquish her ego, and enter a transitory space
beyond the restrictions of temporal and spatial locality. It is in
this shamanistic state that Fern channelled energies into the
land that enveloped her; the fervent raffia blurred the visual,
spatial, and metaphysical boundaries between body, earth, and
air. Othello's camera captured glimpses of the rituals to serve
as memory and provocation for future rituals to continue the
healing through what Joanna Macy describes as the 'Great Turning.'[1]

Fern Shaffer
Painter, performance artist, lecturer and environmental advocate

Fern Shaffer and Othello Anderson, *Nine Rituals*, 1995 - 2003
↑ 3rd Ritual, March 9, 1997 in a cornfield outside of Mineral Point, Wisconsin
↓ 9th Ritual, September 9, 2003 at Cache River basin, Illinois
(p.98) 2nd Ritual, February 9, 1996 at the edge of the Pacific Ocean, Big Sir, California

SCHLEICHERAOLEOSA

Kusum

Schleichera oleosa, or Kusum, grows abundantly in Chhattisgarh, in central India. It fruits in May and June, just before the monsoon. Gonds, one of the indigenous communities residing within central Indian forests, use this edible fruit in a variety of ways. It is a source of food and energy to itinerant herders who travel long distances in search of suitable buyers and satisfactory returns for their cattle. When school halts in summer, children sit under Kusum's shade, bonding over the fruit's sweetly-sour taste, while their livestock graze nearby. Both humans and birds consume its flesh, discarding seeds. Old and young residents collect these discarded seeds daily, accruing large quantities. Grinding these seeds yields oil used year-around by residents for cooking, curing infections and inflammations, and treating joint paints.

Kusum trees are also home and food to *Kerrialacca*, an insect that consumes the tree's sap to produce a resinous substance called lac. Lac is coveted by many industries—cosmetics, food processing, wood-making, and textiles. Collected from forest depths, it is categorized by the government of India as a Non-Timber Forest Produce (NTFP), a livelihood option for Gond communities. Kusum shapes the insect's life-cycle. Only soft and newly-sprouted tree branches allow the insect's proboscis within.

In a Gond village in Kanker district of Chattisgarh, where I have worked since June 2017, not all plants and animals are welcome. *Ficusreligiosa*, or peepal, is a parasite for Kusum, a *parjeevi* whose roots deprive Kusum of nutrients in the soil. A peepal growing nearby can slowly kill Kusum in 15 to 20 years, some say. Squirrels contribute, too, by dispersing peepal seeds widely, and termite mounds manufacture perfect soil for peepal. If a squirrel brings a peepal seed with traces of termite-produced soil to a Kusum branch, peepal grows quickly. Once, this plant represented a place beyond culture—a place inhabited by nature. Now, I see it as a strange hybrid of interspecies assemblages where nature and culture come together to subvert 'modern' schemes. It is these interspecies interactions that my ethnographic research highlights in an otherwise anthropocentric discussion of sustainability, where the consuming *anthropos* is both complicit and accountable. My research suggests that insects, plants, animals, and even physiochemical elements can critically influence production of lac resin, in turn affecting livelihoods. Previously, interventions to improve fluctuating lac yield focused on mono-culturing, growing a single species with pesticides and insecticides while gradually decreasing the diversity and resilience of an ecosystem. Countering this demands granular inspections of roles of and interactions between various actors in the Kusum's ecosystem. Thinking with the Kusum's entangled socio-natures means ethically and responsibly imagining sustainable futures while also engaging in social and political negotiations with our ethnographic interlocutors.

Vinisha Singh Basnet
Research Associate
Centre for Development Practice, Ambedkar University Delhi

Collecting Kusum seeds for oil extraction

Shorea robusta

SHOREA ROBUSTA

Sal (English, Hindi and Bengali)
Sakhua (Sadri)
Makka (Oraon)
Sarjom Daru (Mundari)

The *Sal* or *Sakhua* is a tall deciduous tree growing to a height of up to 50 meters and girth of 1.5 to 2 meters. The leaves are ovate-oblong, and the flowers are yellowish in lax axillaries or terminal panicles. The fruits are indehiscent, ovoid with 5 equal wings. The seeds are ovoid in shape. Found mostly in eastern, central, and northern India, it generally flowers between March and April, and the fruiting season is between June and August.[1]

Apart from a scientific mapping of the plant, it is interesting to understand how the Sal tree is intimately connected to the ethno-cultural and ethno-religious belief system of the indigenous or tribal people of the Indian state of Jarkhand and its adjoining areas. The *Sal* tree is central to the celebration of the *Sarhul* or *Baha Parab* (festival of flowers), a festival dedicated to Mother Nature and celebrated by the aboriginals of Jharkhand and its adjoining areas, chiefly the Munda, Santhal, Kharia, and Ho who are primarily agriculturalist by profession and chiefly follow the Sarna religion. The term 'Sarhul' is derived from two words – 'sar' meaning seed of Sal and 'hul' meaning worship; so Sarhul means the worship of the Sal tree, or Nature. The rituals of Sarhul are performed under the Sal tree in the sacred grove called *sarna sthal*, the abode of the village deity Goddess Sarna. In Sarhul, the Oraon worship Lord Dharmesh and Mother Nature for the blessings they bestow upon all the living forms on earth. Sal tree is also associated with *Phagua* feast. Under the Sal tree the worship of Singbonga, the Supreme Being in the Sarna religion, takes place. The Sal tree is also associated with the marriage ceremony of the indigenous people of Jharkhand. It is believed that unless the bridegroom sits on an altar made of Sal, the marriage will not be recognised. Thus, the Sal tree represents Mother Nature, and various sacred rituals are performed in its worship to thank Mother Nature for offering food, fire, and shelter to humans. Besides offering the basics for human existence, the Sal tree is also useful for its medicinal values. The Mundas use the bark and the seeds against dysentery. The resin of the tree is used as an astringent and as an ingredient for skin diseases.[2] It needs to be argued that this human/nature relationship based on an equal regard and respect toward each other and mediated across centuries through the culture, religion and rituals of the indigenous people of the region serves as one of the many extant examples of how humans and nature still exist in close proximity with each other in rather co-constitutive ways, unlike the dominant Western hegemonic model of human/nature duality. This eco-theological model of human/nature co-existence may serve as a corrective to the ways in which humans have tried to dominate nature and as an answer to the planetary environmental crisis of global warming and climate change in this age of the Anthropocene.

Dr. Animesh Roy
Assistant Professor in English
St. Xavier's College, Simdega

Dr. Emmanuel Barla
Principal and Professor in Political Science
St. Xavier's College, Ranchi

THYMUS VULGARIS

Thyme

The Herb for Nightmares

I might sing of Hymettus, the honey mountain,
the Crazy Mountain, where the bees
are all mad for it, brawling and sprawling
over the pink flower-cushions of thyme.

I might sing of the honey, pouring like amber
into the jars, tasting of flowers and resin,
of pepper, dates, cloves, and smelling
of burned plastic and pencils.

I might sing of its neat green leaves like pins,
its creeping unassuming habit, its familiar
scent and savour, its kitchen-haunting
presence in all we cook with meat.

I might sing of its woody stems, simmered
in water for disinfectant, smelling
of health and cleanliness, piercing
the clog and slow drain of colds.

I might sing of maidens embroidering
the furious black bees on thyme sprays,

whose fierce strength brought ardour
to the knights who wore their favours.

But I will choose to sing of the quiet
comfort thyme brings to those who suffer
with 'phrensie and lethargie,' its peace for those
whose sleep is plagued with nightmares.

Elizabeth Rimmer

The phrase 'phrensie and lethargie' comes from Culpepper's Herbal, and is the earliest reference I have seen to bipolar disorder.

This poem first appeared in *Haggards*, published by Red Squirrel Press 2018.

TOXICODENDRON RADICANS

Poison Ivy

In the dense woods and edge habitats of Eastern North America, *Toxicodendron radicans* thrives. Thick vines wrap around oaks, their leaves of three emerging from the hairy vines to catch the sunlight, or appearing as smaller plants along the disturbed edges of forest paths or lawns. In the early spring in upstate New York, the air is thick with pollen from small flowers, which soon covers every surface with a thin layer of powdery yellow. I'm not allergic to pollen but I am allergic to poison ivy, and my skin hardens into bumps and lines where I have made contact with the potent *urushiol* contained within the plant body. When the rash first arrives it is tenderly itchy, and when I feel it with the skin from my fingers, the swollen skin feels somehow external to my body. It takes 24 to 48 hours for skin to react to the urushiol found in *Toxicodendron radicans*. Prince wrote "You didn't have the decency / to change the sheets." If you don't wash your sheets after coming into contact with poison ivy, you risk continued re-exposure.

Toxicodendron radicans produces urushiol not as a defense mechanism, but as a way to retain water and to protect the plant against fungal infections. Few species other than *Homo sapiens* are allergic to urushiol, even as they eat the leaves, pollinate the flowers, or consume the stone-fruits that form in late fall. For some people, mere contact with poison ivy prompts a reaction known as cell-mediated immune response. Skin cells lose the ability to communicate with other cells as urushiol binds to the proteins in their membranes.[1] Exposing oneself to poison ivy will make the reaction worse over time. Urushiol has a cumulative effect, causing increasingly severe reactions with each instance of exposure. I find it disturbing and incredibly opening to learn that the pain of exposure is not intentional. It matters, and it doesn't matter, that it is not done to hurt me.

In 2006, a study indicated that climate change and rising CO_2 levels cause poison ivy plants to grow larger, and to produce more potent doses of urushiol.[2] Poison ivy arrives and thrives in an atmosphere that is disturbing to, and disturbed by, human cohabitants.

As an individual poison ivy plant responds to increased airborne carbon, I respond to urushiol as it crosses the boundary of my skin. There is an intimacy in this infection, its particular inflection, in the communication that occurs in this plant's own terms. In an interview with Laverne Cox at the New School in 2014, bell hooks poses the prescient question, "what does it mean for us to cultivate together a community that allows for risk – the risk of knowing someone outside your own boundaries?"

Not every encounter is a safe encounter. What does it mean to cultivate a future together with *Toxicodendron radicans*? Can we negotiate a multispecies cohabitation with this strange bedfellow, whose chemical compounds permeate our sheets, whose leaves of three threaten me? Leaves of three, enter me.

Lindsey french
Visiting Assistant Professor
Department of Studio Arts, University of Pittsburgh

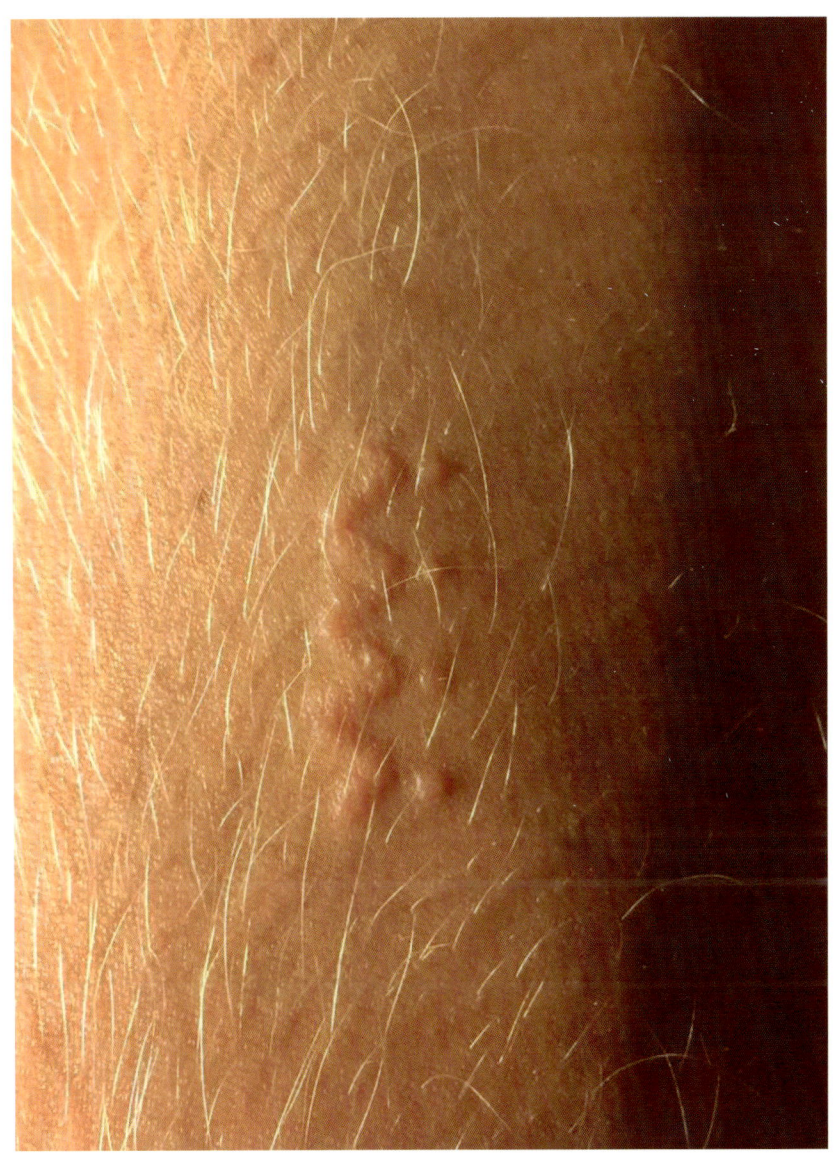

Lindsey french, *urushiol tattoo*, 2016

TSUGA CANADENSIS

Eastern Hemlock

The genus "Tsuga" is from the Old Japanese 栂 (tsuga or toga) formed from the characters for tree 木 + mother 母. These "mother trees" from Canada ("canadensis") are a species of evergreen tree.

The Redwood of the East

Not to be confused with the hemlock of Socrates, Eastern Hemlock is a long-living evergreen tree replete with tiny cones and flattened, white-striped needles; the ecological foundation of its eponymous forests and for more North American forest types than any other tree; home to unique assemblages of birds and beasts, ants and spiders, fungi and bacteria; shady and cool in summer, warm deer-bed in winter; storehouse for the carbon we pump into the atmosphere; cultural touchstone for many New England poets.

Eastern Hemlock grows throughout eastern North America, wherever people are or are not: along rivers and streams; in valleys; on north-facing slopes; in northern mixed woodlands; in front and back yards. It lives for centuries, attains 50 meters in height, at least 2 meters in girth. Its needles decompose slowly, yielding thick spongy organic soils that support its seedlings and silence our footsteps.

Eastern Hemlock is a source of tanbark for treating leather and timber for post-and-beam constructions. Astringent needles and bark are decocted for kidney stones, rheumatism, fevers, colds, coughs, diarrhoea, and scurvy. Its inner bark—stripped, sliced, soaked—is edible in a pinch.

Slowly Vanishing

Eastern Hemlock is slowly vanishing at the mouths of insects and the hands of wo/men. Since the glaciers receded 14,000 years ago, Eastern Hemlock has been one of the most common trees east of the Mississippi River. Now, it is being slowly sucked dry by the hemlock woolly adelgid, a small mealy insect introduced into eastern North America by the horticulture industry in the early 1950s. On the wind, the wings of birds, and the cars of wo/man, the adelgid has been carried north, south, east, and west—settling, feeding, breeding, and moving on, leaving disintegrating *Tsuga* skeletons in its wake, unless pre-emptively salvaged for biofuel chips and occasionally lumber.

Be the Tree
Every second: breathe in, breathe out
For minutes: stand still
For hours: be silent
For days: be aware
For years: live rooted
Over centuries: watch and wait.

Aaron M. Ellison
Senior Research Fellow in Ecology
Harvard Forest and Department of Organismic & Evolutionary Biology
Harvard University

David Buckley Borden, Aaron M. Ellison & Salua Rivero, *Wood Shoes*, 2017
0.3×0.7×0.7 meters, wood and acrylic paint;
a part of the Hemlock Hospice installation at the Harvard Forest
(p.130)

URTICA DIOICA

Stinging Nettle

Stinging Nettle, *Urtica dioica*, is native to parts of North America, Europe, North Africa, and Asia and is widely used as a fibre, medicine, and food plant. It is an herbaceous perennial that grows two to four feet tall, has toothed, opposite leaves, and small green or brown flowers that grow in dangling clumps. Although a single plant produces hundreds of seeds, it largely spreads rhizomatically, preferring to live in dense patches. Stinging nettle is not generally considered a beautiful plant, but it is a nourishing and healing plant that helps to restore balance to ecosystems. Perhaps its beauty lies in the relationship it has with other beings.

Stinging nettle is a dynamic accumulator and an ally to at-risk plants, healing soil that has been damaged by decades of capitalist-industrial agriculture and seemingly insatiable urban growth. In polluted areas, stinging nettle helps reduce toxins in the soil by drawing them through its roots up into its leaves. When grown in nutrient rich soils, its decaying leaves act as fertilizer for other plants.

The "sting" of the nettle comes from the irritating chemicals that are released through hairs on its stem and the underside of its leaves. Momentarily painful, the sting causes a numbing sensation at the site for hours to days. Some people claim, much like bee stings, that repeated stings from the nettle can diminish instances of arthritis and other joint problems.

Stinging nettle grows abundantly, wildly, and is only rarely cultivated. It is foraged by people who have discovered beyond its sting that its ally-ship extends to humans. Young nettle leaves are carefully picked with gloved hands to be used in pesto, soups, curries, and savoury pastries. Stinging nettle is a nutrient-packed plant, high in iron, calcium, vitamins C and A, and rich fatty acids. Medicinally, stinging nettle is used for a wide variety of ailments. It is a diuretic and is useful for treating urinary tract infections and the early stages of enlarged prostates. It is used as an herbal remedy to treat gout, diabetes, and hay fever. As a general tonic, stinging nettle infusions and teas restore strength and vitality in stressed bodies. For centuries, the fibres of its stem were transformed into a coarse fabric used for clothes, rope, and bags.

Some subspecies of stinging nettle contain formic acid, one of the main treatments used to treat bee colonies that are infested with Varroa mites—perhaps bees living close to nettle patches have already discovered this without the knowledge of their beekeepers.

Whether emerging along polluted rivers, stinging the hands of people suffering from arthritis, providing nutrients to the bodies of depleted, stressed people, or being chewed by mite-ridden worker bees, stinging nettle certainly seems to know which living bodies—soil, people, bees—need healing.

Rebecca Ellis
Ph.D. Candidate, Geography, University of Western Ontario
Resident Member, Rotman Institute of Philosophy

Urtica-Spinach with eggs and potato

Stills from ethnographic film, *Yuyos*, 2018
Michał Krawczyk,
Ph.D. Candidate, Environmental Humanities, Griffith University

V E R N O N I A TWEEDIANA

Jagua pety

In Paraguay, *yuyos* are spontaneous plants and herbs that can be taken for their curative properties. Also called *pohã ñana*, which in the Guarani language means medicinal plant, they are essential to the practices of natural medicine. In everyday life, they are commonly consumed through *tereré*: a social beverage of chilled water infused with yuyos poured in a vessel of *yerba mate*. Twice a day, this ritual drink is shared among people passing on knowledge and stories of human-vegetal communion. Nowadays, as deforestation devours much of the Paraguayan territories to make room for industrial monocultures, genetically modified plants are substituting the wild ones and changing the natural cultural landscape. In the remaining forests, countless yuyos are hiding away to emerge only before the eyes of those who can see. One of them is Jagua pety[1]: an herb that still grows in Colonia Luz Bella, where a peasant family of agroecologists is dedicated to protecting the local environment. Moved by the desire to narrate a story about their ethnobotanical knowledge[2], my partner and I immersed ourselves in the lives of Doña Emeteria and Don Franco's family. As the human relationships grew more intimate, eased by the trusting conviviality of tereré, we gradually acknowledged a more-than-human world where yuyos also enacted their role for the household wellbeing. In their curative action, they became co-narrators of an ecology struggling to (re)exist, spurring the human others to tell their personal stories of engagement and eco-resistance. Jagua pety was one such narrating actor. Our interaction began as we noticed it drying, hung in the porch near our accommodation. It had let Doña Emeteria gather its leaves to prepare a tincture for the forthcoming winter. We were going to do it together, slowly, with patience.

One day to sort the leaves out.
One day to place the leaves in a dark jar to be left in darkness.
Another day to complete the tincture: one more winter.

As time went by Jagua pety helped us talk and fantasise, building up imagination for the moment of its medicinal transformation. Meanwhile, we discovered that around Colonia Luz Bella Jagua pety is as rare and famous as Doña Emeteria's abilities with natural medicine. It is a great expectorant, and she is one of the few women who still make its tincture. People from far away come to treat their cough, for it is said that despite industrial medicine, Jagua pety remains their favourite doctor, and at times their best vet, as the herb is also given to sick chicks to cure infections [3].
In Paraguay, Jagua pety is going down with the trees. Yet, it keeps being the nonhuman hero of its own story because an agro-ecological family is resisting deforestation by becoming one with the vegetal others. This is their story of shared perseverance.

For 100gr of dried yuyo, use 260cl of water and 240cl of 96° alcohol. First the water, then add the alcohol, and wait a little. After, pour on the yuyo, and leave it for 15-20 days. Once filtered, adults take 20-25 drops in a glass of water twice a day and children 5-8 drops.

Giulia Lepori
Ph.D. Candidate, Environmental Humanities
Griffith University

ZEA

MAYS

Corn
Maize

Pesticide I: Calypso
"... from καλύπτω (kalyptō), meaning 'to cover', 'to
conceal', 'to hide', or 'to deceive'."

When at last he reaches
151 South, stunning corn
fields flank the road,
emerald sleeves rippling out
in every direction. In the heat
of mid-July, the corn is always
undressing––her silk undone,
her layers falling away. Each ear
like a shuttle of gold.
Untroubled by corn borers
or worms or aphids, and without
a weed between the rows,
this corn is the goddess
of her landscape. Nothing
thrives here but corn.

Mesmerizing, she lures
the farmer who thinks only of the
comfort, the ease.
No real way of knowing
what enslavement lies ahead.
While small rectangular labels
at the edges of the fields
quietly reveal the poisons,
he does not pay attention
to the signs.

Heather Swan

"Caylpso" is part of a group of poems which take as their titles words that chemical
companies have used to name their pesticides.

Li Lisha (李丽莎), *The family of the Aunty L* (六姑一家子), 2006

CORNSTITUTION AND ANTHERMATION OF THE RIGHTS OF MAIZE

Excerpts

Herein, the Cornstitution translated into human language the already organized governance of the corn. The Cornstitution is like articles of governing. This Cornstitution describes the political constitution, regulation, and governing of the constituent relations that are maize.

Corn pollinates, germinates and fruits the rights and responsibilities of different parts of its governable bodies. The Cornstitution also anthermates the rules or methods of relating between maize and corn parts and to foreign parts. Look at the corn dance, eat a fresh cob, you can see, you can taste, you can feel, corn has a sense of humor.

The Cornstitution affirms and asserts the inherent and germinated rights and responsibilities of corn and its parts, holes, wholes, partialities and communities. Rights circulate from the being of the corn and are conferred upon it by the already organized and dynamic governance of the corn. Rights rest in the kernel, the stalk, the silk, the roots, anthers, and leaves and every part of the corn at all scales.

Rights are matched, completed and leavened by Responsibilities that inhere at every scale, including the extended cornsciousness and temporalities of the corn bodies. These rights and responsibilities are to be enacted by and granted to patches, kernels, cobs, stalks, silks, roots, anthers, and leaves of every different size, colour; yellow, white, purple, blue, magenta, red, black, and rainbow; and variety; sweet, feed, polenta, pop, and hominy.

Each patch, kernel, stalk, cob, silk tassel, root, anther, and leaf has rights to be supported by and the right to their respective patches, kernels, cobs, stalks, silks, roots, anthers, and leaves.

Corn bred: The Cornstitution affirms the inalienable, inherent and corngenital rights of sex, free pollination and horizontal gene transfer. Access to sunlight, water and air is a reserved and protected right for all maize. As is the right to share sweetness. And the right to be corny.

The right to cornsort with helpers, and the right to be communicated by other mediums. The right to use sweetness to make deals with carriers and mediums. Corn reserves the right to determine place and to remain isolated in hill, valley and dale. This includes the right to resist cross-pollination or gene-splicing and the right to grow without wind and to not be forced into international or inter-valley commerce. Additionally, each kernel or cob may enlist any other kind of being into spreading and propagating corns in so far and as long as that being won't violate or be detrimental to the rights and responsibilities of any and every patch, stalk, cob, kernel and gene that might come out of the enlisting or recruiting of that being.

Defense — when the rights and responsibilities of
corn are threatened by foreign parties, corn has an
obligation to use its sweetness, or its sharpness or
any other means integral to its body or spirit, in order
to defend maize collectives through defensive and
offensive means including but not limited to diabetes,
heart disease, stabbing, corns and bunions, tooth
decay and crop failure.

The Cornstitution emerged from a desire to commemorate
indigenous presences and jurisprudence on land that has
been marked by new political jurisdictions such as "Illinois"
and "United States." On October 12, 2009, the contested
USA holiday called Columbus Day, duskin drum, Sarah
Lewison and Wendy Weinhold convened a meeting with
a cornfield. We had been researching the implications
of gene seizure and contamination by GMO technologies
and wanted to radically deprecate anthropomorphic/
human expectations of knowledge and communication
transfer, by attempting to listen to the corn. A great deal of
brainstorming ensued and the Cornstitution was brought to us.

Sarah Lewison & duskin drum

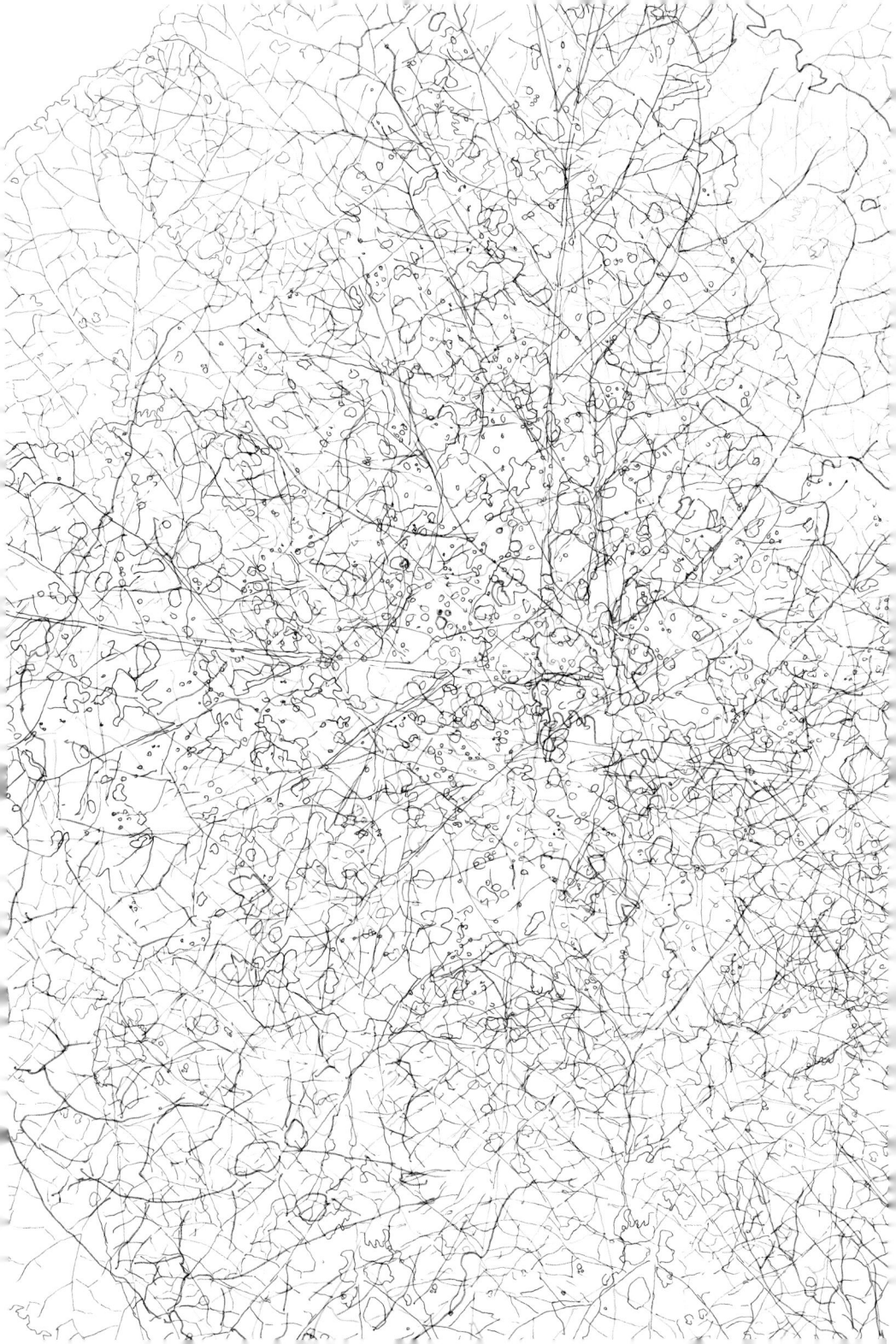

E N D N O T E S

Editors' Note

1. Concept of *becoming* from philosphies of Deleuze and Guattari, on *molecule* and *becoming* see: Gilles Deleuze and Félix Guattari, *A Thousand Plateaus*, Brian Massumi, trans., (London, UK: Continuum, 2004), 300-304.

2. John D. Key, *Chinese Herbs, Their Botany, Chemistry, and Pharmacodynamics* (Rutland, Vermont: Charles E. Tuttle Company, 1976), 9-10.

Amaranthus

1. "Amaranth grain, cooked", SELF Nutrition Data, http://nutritiondata.self.com/facts/cereal-grains-and-pasta/10640/22.

2. Chola Con Cello, "Foods of the Americas: Amaranth, the Outlaw Grain," Medium, https://medium.com/@cholaconcello/foods-of-the-americas-amaranth-the-outlaw-grain-946957d9e51.

3. Daniel K. Early, "Amaranth Production in Mexico and Peru," *Advances in New Crops*, J. Janick and J.E. Simon, eds. (Timber Press, Portland, OR, 1990), https://hort.purdue.edu/newcrop/proceedings1990/V1-140.html.

4. "Assembly resolution on mining against SC order: Panel to MPs," *The Times of India*, 12 August 2018, https://timesofindia.indiatimes.com/city/goa/assembly-resolution-on-mining-against-sc-order-panel-to-mps/articleshow/65370566.cms.

Artemisia vulgaris—Mcgwyrt

1. **Lesley-Caron Veater** is a transpersonal therapist working with dreamwork, art and art journaling and has been an herbalist and permaculture gardener for more than 30 years. Within Lesley's practice, Windhorse Dreamwork, a unique range of workshops are offered with a focus on creativity and dreamwork.

2. **Professor Brian Bates** was Chairman of Psychology at the University of Sussex and creator of the course there in Shamanic Consciousness. He recovered the ancient way of magical life called 'Wyrd' from Anglo-Saxon history and has reintroduced it into modern culture through his books and courses, especially *The Way of Wyrd* and *The Real Middle-Earth*.

3. References: Balds Leechbook, in *Leechdoms, Wortcunning, and Starcraft of Early England*, ed. O. Cockayne, Rolls Series, 3 vols (London: Longmans, 1864-66), vol II (1864).

   Brian Bates, *The Way of Wyrd*, (London: Century, 1983; Hay House, 2012).

   Brian Bates, *The Real Middle Earth*, (London: Macmillan, 2002).

   Brian Bates, *The Wisdom of the Wyrd*, (London: Random House, 1996).

Basidiomycota biocompositum

1. G. A. Holt, et al., "Fungal Mycelium and Cotton Plant Materials in the Manufacture of Biodegradable Molded Packaging Material: Evaluation Study of Select Blends of Cotton Byproducts," Journal of Biobased Materials and Bioenergy 6, no. 4 (2012): 431-439.

Carpobrotus glaucescens

1. Margaret Iselin et al., *Jandai Language Dictionary: A Dictionary of Language Spoken on Stradbroke and Moreton Islands Based on Words Remembered by All Elders and Recorded by Interested Visitors to Our Shores* (Dunwich, Qld: Minjerribah Moorgumpin Elders-in-Council Aboriginal Corporation, 2011).

2. Renata Buziak, "Biochromes: Perceptions of Australian Medicinal Plants through Experimental Photography" (PhD diss., Queensland College of Art, Griffith University, 2015).

Gymnocladus dioicus

1. "Plant Guide: KENTUCKY COFFEETREE Gymnocladus dioicus (L.) K. Koch," USDA Plant Guide, http://plants.usda.gov/plantguide/pdf/cs_gydi.pdf.

2. Daniel H. Janzen and Paul S. Martin, "Neotropical anachronisms: the fruits the gomphotheres ate," *Science* Vol. 215, Issue 4528 (1 Jan 1982): 19-27, https://doi.org/10.1126/science.215.4528.19.
3. Connie Barlow, "Anachronistic fruits and the ghosts who haunt them," Arnoldia 61, no. 2 (2001): 14-21.
4. Melvin R. Gilmore, *Uses of Plants by the Indians of the Missouri River Region*, SI-BAE Annual Report #33 (Washington Government Printing Office, 1919).
5. "Plant Guide."
6. Missouri Botanical Garden, http://www.missouribotanicalgarden.org.

Helianthus tuberosus

1. Arthur Parker, *Iroquois Uses of Maize and other Food Plants* (Albany, NY: State Museum Bulletin No. 144, 1910), 106.
2. Carol Cornelius, *Iroquois Corn in a Culture-based Curriculum* (Albany, NY: SUNY Press, 1999), 94.
3. Gloria Anzaldúa, *Borderlands – La Frontera, the New Mestiza* (San Francisco, CA: Aunt Lute, 1987), 49.

Mors ontologica

1. Timothy Morton, *Dark Ecology: For a Logic of Future Coexistence* (New York: Columbia University Press, 2016), 55.
2. Philip K. Dick, *A Scanner Darkly* (New York: Mariner Books, 2011), 264.
3. Dick, *A Scanner Darkly*, 284.
4. Ibid.
5. Morton, *Dark Ecology*, 56.

Oryza sativa

1. P. K. Gode, "History of the Rangavalli (RĀṄGOḼĪ) Art - Between CAD 50 and 1900," *Annals of the Bhandarkar Oriental Research Institute* 28, no. 3/4 (1947): 226-246.
2. Jyotsna S. Kilambi, "Toward an understanding of the Muggu: Threshold drawings in Hyderabad," *RES: Anthropology and Aesthetics* 10, no. 1 (1985): 71-102.

3. Anna Laine, "In conversation with the Kolam practice: auspiciousness and artistic experiences among women in Tamilnadu, South India" (PhD thesis, School of Global Studies, Social Anthropology, University of Gothenburg, 2009).
4. Gode, "History of the Rangavalli," 226-246.
5. Marie Rasmussen, "In Search of Lakshmi's Footprints: A Brief Study of the Use of Surface Design in India. Fulbright-Hays Summer Seminars Abroad, 1997 (India)" (United States Educational Foundation in India, 1997).
6. Martyn Rix. *The Golden Age of Botanical Art* (University of Chicago Press, 2013).

Picea glauca

1. Waasebines wrote and made an audio recording (with me) of the Ojibwe language portion of this teaching. Slight spelling changes were made by Wendy Geniusz, as was the implied translation.
2. Most likely, Asemaa was originally *Nicotina rustica,* a different species of tobacco than the plant used in the production of commercial cigarettes (M. Geniusz 152). Towards the end of his life, Waasebines often spoke of the problems of making offerings with store bought tobacco, which caused so many health problems in Indigenous communities.
3. For more information: on Creation see W. Geniusz, *Our Knowledge,* 57; on offerings see W. Geniusz, *Our Knowledge,* 60-63.
4. References: Wendy Makoons Geniusz, *Our Knowledge is Not Primitive: Decolonizing Botanical Anishinaabe Teachings* (Syracuse: Syracuse University Press, 2009).
   Mary Siisip Geniusz, *Plants Have So Much to Give Us, All We Have to do is Ask: Anishinaabe Botanical Teaching,* ed. Wendy Makoons Geniusz (Minneapolis: University of Minnesota Press, 2015).

Plantago major

1. Younes Najafian et al., "Plantago major in Traditional Persian Medicine and modern phytotherapy: a narrative review," *Electronic Physician,* 10(2) (February 2018): 6390–6399. The authors studied uses of

Plantago major within these traditional texts and found that many of them could be confirmed with modern medical research.

2. "White Man's Foot: Broadleaf Plantain," Weed Science Society of America, accessed 2018, http://wssa.net/wp-content/themes/WSSA/WorldOfWeeds/whitemansfoot.html.

3. Karen Louise Jolly, *Popular Religion in Late Saxon England: Elf Charms in Context* (Chapel Hill: University of North Carolina Press, 1996), 125.

4. Robin Wall Kimmerer, *Braiding Sweetgrass: Indigenous Wisdom, Scientific Knowledge, and the Teachings of Plants* (Minneapolis: Milkweed Editions, 2013).

5. Kimmerer, *Braiding Sweetgrass*, 214.

Populus deltoides

1. Quoted in Joseph Epes Brown, *The Sacred Pipe: Black Elk's Account of the Seven Rites of the Oglala Sioux* (University of Oklahoma Press, 1953), 75.

2. Robin Ridington and In'aska Dennis Hastings, *Blessing for a Long Time: The Sacred Pole of the Omaha Tribe* (University of Nebraska Press, 1997), 52.

Pueraria montana var. lobate

1. Dr. Zuleyka Zevallos, "What is Otherness," Other Sociologist, https://othersociologist.com/otherness-resources.

2. Amanda Taub and Max Fisher, "In U.S. and Europe, Migration Conflict Points to Deeper Political Problems," *New York Times*, June 30, 2018, https://www.nytimes.com/2018/06/29/world/europe/us-migrant-crisis.html.

3. "Glossary of Invasion Biology Terms", Wikipedia, https://en.wikipedia.org/wiki/Glossary_of_invasion_biology_terms#Terminology.

4. Craig Welch, "Half of All Species Are on the Move—And We're Feeling It: As climate change displaces everything from moose to microbes, it's affecting human foods, businesses, and diseases," *National Geographic*, April 2017, https://news.nationalgeographic.com/2017/04/climate-change-species-migration-disease.

5. Bill Finch, "The True Story of Kudzu, the Vine That Never Truly Ate the South," *Smithsonian Magazine*, September 2015, https://www.smithsonianmag.com/science-nature/true-story-kudzu-vine-ate-south-180956325.

6. Jonathan Foley, "It's Time to Rethink America's Corn System," *Scientific American*, March 5, 2013, https://www.scientificamerican.com/article/time-to-rethink-corn.

7. Timothy Lee Scott, *Invasive Plant Medicine: The Ecological Benefits and Healing Abilities of Invasives* (Rochester, Vermont; Toronto, Canada; Healing Arts Press, 2010), 237-42.

Punica Granatum

1. John M Riddle, *Eve's Herb's: A History of Contraception and Abortion in the West* (Cambridge: Harvard University Press, 1997).

2. Suzanne Raga, "9 Forms of Birth Control Used in the Ancient World," accessed March 26, 2019, http://mentalfloss.com/article/83685/9-forms-birth-control-used-ancient-world.

3. Gina Kolata, "In Ancient Times, Flowers and Fennel for Family Planning." *The New York Times*, August 1994.

Raphia

1. Joanna Macy and Chris Johnstone, "The Great Turning (adapted from Ch.1)," Active Hope, accessed 2019, https://www.activehope.info/great-turning.html.

Shorea robusta

1. David Prain, *Bengal Plants*, vol. 1 (Calcutta: Gouranga Das Press Pvt. Ltd., 1963).

2. P.P. Hembrom, *Ethnomedicine* (Ranchi: I.D Publishing, 2011).
    P.P. Hembrom, *Indigenous Medicine*, 7 vols., (Ranchi: Catholic Press, 2011).
    P.P. Hembrom, *Indigenous Medicine for Life* (Ranchi: I.D Publishing, 2010).

Toxicodendron radicans

1. John Barrat, "A Poison Ivy Primer." *Torch, Smithsonian,* August 12, 2014, https://torch.si.edu/2014/08/a-poison-ivy-primer/.
2. J. E. Mohan et al. "Biomass and Toxicity Responses of Poison Ivy (Toxicodendron Radicans) to Elevated Atmospheric CO2." Proceedings of the National Academy of Sciences, 103, no. 24 (2006): 9086–9089, doi:10.1073/pnas.0602392103.

Vernonia tweediana

1. In Guarani, the second official language of Paraguay, Jagua pety means "the dog's tobacco".
2. Their story is recounted audio-visually in 'Yuyos' (2018), the first feature ethnographic film - independent and self-produced, by Michał Krawczyk and Giulia Lepori. Trailer and information at echoesofecologies.noblogs.org/yuyos/.
3. Its root is mashed and put in their drinking water, mostly to cure flu.

PHOTOGRAPHS ARTWORKS

Cover, p.iii, p.xv-xvi, p.126: *Brain Leaf* ©2016 Nathalie Lavoie

p.viii: ©2018 josh Armstrong

p.7: *Destroying Angel* ©2018 Emily Arthur

p.8: *Amaranthus* ©2017 Zuri Camille de Souza

p.13: *Yellow Crushing argan seeds* is a derivative of *Crushing argan seeds*, 2013 by Giulia Corradini, licensed under CC BY-SA 4.0

p.16: ©Andrea Haenggi

p.20: *Material by Ecovative*, 2012 by Stephen P Nock, licensed under CC BY SA 3.0

p.24-26: *Carpobrotus glaucescens* ©Renata Buziak